Climate-Challenged Society

Climate-Challenged Society

By

John S. Dryzek, Richard B. Norgaard, and
David Schlosberg

OXFORD
UNIVERSITY PRESS

OXFORD
UNIVERSITY PRESS

Great Clarendon Street, Oxford, OX2 6DP,
United Kingdom

Oxford University Press is a department of the University of Oxford.
It furthers the University's objective of excellence in research, scholarship,
and education by publishing worldwide. Oxford is a registered trade mark of
Oxford University Press in the UK and in certain other countries

© John S. Dryzek, Richard B. Norgaard, and David Schlosberg 2013

The moral rights of the authors have been asserted

First Edition published in 2013

Impression: 1

Published in the United States of America by Oxford University Press
198 Madison Avenue, New York, NY 10016, United States of America

British Library Cataloguing in Publication Data

Data available

Library of Congress Control Number: 2013941633

ISBN 978–0–19–966010–0 (Hbk.)
ISBN 978–0–19–966011–7 (Pbk.)

Printed and bound in Great Britain by
CPI Group (UK) Ltd, Croydon, CR0 4YY

Preface

We offer here a short and, we hope, clear, readable, and accessible book about the challenges that climate change presents, and about how societies might respond. It is not a book about the science of climate change, which we take as more or less given. Nor is it just a survey of the work of others, though we do build upon what others have done. This book is intended rather as a critical and integrative introduction, written "with attitude," because we do not think that anyone else has yet really come to grips with the full range of relevant issues, or set out the pieces of the puzzle in effective fashion. While we do our best to fit the pieces together, the magnitude and inevitable persistence of the climate challenge means that no once-and-for-all solution identified by us, or anyone else, is attainable. We synthesize and deploy cutting-edge scholarship on climate change and society in ways we hope provide the basis for intelligent thought and collective action.

The biggest challenges prove to lie not in the science or technical aspects of climate change itself, but rather in social, economic, and political systems as they interact with natural systems: be it in creating climate change, trying to prevent it, or coping with its effects. Climate change presents us with a mirror in which we can see the best and worst of human society. If currently the worst seems to have the upper hand, it does not have to stay that way.

All three of us have worked on environmental issues, including but not limited to climate change, for several decades. We co-edited *The Oxford Handbook of Climate Change and Society*

(Oxford University Press, 2011), and this book is in some ways an outgrowth of that large effort. The *Handbook* featured contributions from 69 of the best people we could find to write chapters on 47 particular topics. What you now have in your hand (or on your screen) is much shorter. While we cite many *Handbook* chapters, this book is more than just a summary of that much longer collection. As we have already noted, we try here to put the pieces together (from the *Handbook* and elsewhere), which has led us in significant new directions. We owe a lot to those 69 authors, and to many others we have worked with on related issues.

John Dryzek would like to acknowledge in particular Hayley Stevenson, with whom he has worked on the global governance of climate change, and Simon Niemeyer, on the deliberative aspects. He would also like to acknowledge the importance of the associations he has developed as a member of the Science Committee of the International Human Dimensions Program on Global Environmental Change, the Earth System Governance project, and the Climate Change Institute at the Australian National University.

Richard Norgaard would like to acknowledge the opportunities he has had to acquire deeper insight through four decades of working with faculty and graduate students in the Energy and Resources Program at the University of California, Berkeley and through his current service on the Intergovernmental Panel on Climate Change, on climate advisory committees of Tsinghua and Beijing Normal Universities, and the Delta Independent Science Board of the State of California.

David Schlosberg would like to acknowledge colleagues at the University of Sydney, in particular those in the Sydney Network on Climate Change and Society and the Institute for Democracy and Human Rights, for the warm and stimulating welcome to a new environment—where thinking about the challenges of climate change is, ironically, enjoyable. Thanks also are due to longtime colleagues and friends in environmental political thought in the US, UK, and Australia for sharing their knowledge and critiques.

We also agree we need to thank each other—it really has been a thoroughly enlightening process producing both of these books over the last five years. We have not always agreed, but the disagreements have always been fruitful.

As always it has been a true pleasure working with Dominic Byatt, our editor at Oxford University Press, and his team.

Contents

List of Figures

List of Acroynms and Abbreviations

CDM	Clean Development Mechanism
CFC	chlorofluorocarbon
CO_2	carbon dioxide
COP	Conference of the Parties
EU	European Union
IPCC	Intergovernmental Panel on Climate Change
NASA	National Aeronautical Space Administration
NGO	non-governmental organization
NPV	net present value
ppm	parts per million
REDD	Reducing Emissions from Deforestation and Forest Degradation
UNFCCC	United Nations Framework Convention on Climate Change
UK	United Kingdom
US	United States

1

Climate's Challenges

Climate change presents a particularly tough challenge. There are of course plenty of other tough problems around: inequality, poverty, terrorism, the instabilities of financial systems, the risks of nuclear technologies, persistent and potentially violent antagonisms in international politics. Yet we have at least a sense of the nature of these sorts of problems, and governments know more or less how to respond to them, although they do not always do so—or do so successfully. The challenges of climate change, by contrast, almost seem to defy comprehension, let alone appropriate response. The science is complex and long contested by a well-financed movement that accuses climate scientists of falsifying conclusions in support of a left-wing political agenda. There is a collision between what the science implies and seemingly common-sense understandings based on casual observation of the weather. The size of the threat calls into question received ideas about the inevitability of human progress: if progress requires continued economic growth based on ever-increasing emissions of greenhouse gases, then that kind of progress is clearly no longer sustainable. The economic stakes could not be higher, calling into question the future of industries such as coal and cars, and leading to deep political conflicts as those whose industries, profits, employment, and lifestyles feel threatened resist the necessary changes.

Even among those who accept the science and recognize the size of the challenge, key questions are hotly debated. So there is dispute about how to think of the risks that climate change brings: for example, should we spend money on preparing for low-probability but potentially catastrophic impacts (such as rapid sea-level rise)? To what extent should we care about the risks our current emissions are imposing on future generations, who will suffer the impacts? If it is accepted that emissions need to be reduced, how rapidly should that reduction occur in light of the economic costs that it will necessarily impose? And how should the burden of reductions be allocated between rich and poor countries, between rich and poor people within countries, between different industries and economic sectors? Does equity here mean that everyone should play their part in reducing emissions, or that those whose societies have built their wealth on a history of fossil fuel use have a much larger responsibility to cut back? If the world needs to move to a new kind of low-emission economy, what should that new economy look like? The range of potential responses is large, and particular actions (such as wind farms, or taxes on fossil fuels) often controversial. Climate change interacts with other sorts of global environmental change (such as biodiversity loss, and ocean acidification), to the extent we are entering the uncharted territory of Earth systems whose trajectory is largely determined by the unintended effects of what people do. National governments and global governance often seem paralyzed in response.

In this book we sort through these issues with a view to how societies might do better. The following questions orient our inquiry, and we base one chapter on each:

- Can climate scientists, publics, and policy makers reach understandings that are sufficiently complementary that they lead to effective action?
- How should we think about assessing and valuing the expected damage of climate change, especially when it stretches out into a distant future, as a basis for deciding in broad terms how to act?

- What are the best kinds of policies and practices for controlling or otherwise mitigating climate-changing emissions, and for effective adaptation to the consequences of climate change?
- How can we seek social justice in the allocation of the benefits and burdens of mitigation and adaptation, especially in light of the fact that those likely to suffer most from climate change are not those most responsible for causing it?
- Can governments and the processes of global governance respond effectively, and if not do we need new forms of governance?
- Can we find ways to cope with an ever-changing "Anthropocene," where Earth systems are highly affected by the mostly negative and unintended consequences of human action?
- What kinds of conceptual shifts, economic ideas, social movement practices, and governance structures might help us transition effectively in response to the profound challenges presented by climate change?

These questions are starting points. As we shall see in subsequent chapters, thinking about them both alone and in combination leads us to recast questions, and to frame new ones.

THINKING ABOUT TIME AND PROGRESS

Accepting the science of climate change as caused by the buildup of greenhouse gases in the atmosphere, as we do, the following seems to be required:

> In order to avoid significant risks, global society needs to reduce rapidly the net emissions of greenhouse gases while also assuring that the burden of the transition to a new economy is borne equitably by the people of the world.

If the world did this, what might life be like in 2050?

Economies would be based on solar, wind, and other renewable technologies, rather than fossil fuels. They would have

3

moved beyond material flows that require constantly throwing things away (perhaps recycling a portion), to an economy of fewer but more durable goods. While poor people would still be attending to basic material needs, the relatively rich would find happiness in cultural, social, and intellectual activities that consume little energy, no longer establishing their identity by the stuff they own and consume (that is, they would be post-materialists).

In spite of these responses, climate change would still be taking place, driven by the historic build-up of greenhouse gases in the atmosphere, which will have been significantly slowed but still far from being in decline. People would be continually adapting to the further dynamics of climate change (some would be migrating). Coastal communities would be dealing with the half-meter rise in sea level while anticipating up to another meter before sea level stabilizes in 2250 or so. Farmers would have developed strategies to cope with more frequent extreme weather and increasing temperatures. Biologists would be assisting critical species to help them move with the changing climate or otherwise adapt to avoid extinction. In spite of an abundance of immediate problems, people would also be looking 50 to 100 years ahead and crafting their current efforts so that each generation will be thankful for the decisions made by prior generations.

Envisioning an appropriate set of responses once the challenge has been defined is one thing, pursuing it quite another. Can established industrial economies and newly industrializing societies really transition to renewable energy technologies? Do individuals, organizations, corporations, and governments really have the necessary courage to battle the fossil fuel interests that try desperately to keep old practices and policies in place? Can people construct new ways to be happy that are less energy and material intensive, and do not require ever-increasing consumption? Can governments wean themselves from their commitment to spiraling economic growth? Can markets serve the general good, not just the self-interest of investors, bankers, and executives? Can collective purpose be re-valued after decades in which government has been disparaged as the antithesis of freedom and choice (especially, but not only, in the United States)? Can scientists and the organizational processes through which people work together handle a rapidly changing environment?

Climate change is not, then, just a scientific challenge requiring a policy response. Also at stake are how people think about meaningful life, how we understand the nature of societies, economies, and governments. These challenges merit serious discussion and analysis, yet early twenty-first century political discourse remains mired in sound bites, attack journalism, short media attention spans, personal attacks (especially in the blogosphere), the manipulation of collective problems for short-term partisan advantage, and a general inability to see beyond immediate material interests.

Especially in Western societies, climate change challenges deep-rooted beliefs in progress. Judeo-Christian religious traditions have long valued the possibility of people transcending their earthly, bodily desires and becoming higher, more moral beings, more like angels, nearer to God. Early in the seventeenth century, Francis Bacon saw science as providing a parallel means of ascent enabling people to control their environment to better meet material needs. Economic, moral, and scientific advance could be mutually reinforcing. Material abundance would reduce crime and the need to exploit others. People would not have to work so hard, leaving more time for moral improvement. By the nineteenth century, progress seemed to entail a grand industrial transformation of raw materials into magnificent infrastructure and prodigious levels of consumption, trends that continue to this day.

Climate change challenges this dream of human progress. But it was not simply a dream; it was a dream that was implemented. People really did transform nature, build more comfortable houses and public amenities, live longer and in many ways better. Climate change now reveals that the systems constructed to further human progress do not fit natural realities. Fossil fuel-based technologies and the infrastructure built upon them are dragged back into the Earthly world that Westerners and others who brought progress thought they were rising above. A dream turned nightmare is difficult to handle. The idea of progress is enmeshed in modern political discourse and the way individuals think about their own future and that of their children. Climate change challenges people to rethink this whole narrative, and that is deeply disturbing.

WHICH SOCIETY, WHICH WE?

Society is climate-challenged, but we might want to think more about the idea of "society." Which society do we mean? Is it national or global? Is it human society in general, or just particular kinds of societies? Is it some aspects of society (such as the economic system, the political system, or the culture), or all of them?

In the context of climate change, society has to mean all of these things. But that does not mean they all merit equal attention. Climate change is first and foremost a global issue—and that is one of the reasons it is proving so intractable. But its effects vary substantially by where people live, and who they are (rich or poor). Currently, it looks as though sub-Saharan Africa and low-lying islands stand to take especially big hits, though river deltas (composing for example much of the Netherlands and Bangladesh), the Arctic, and South Asia should the monsoon move into a drier pattern, are also extremely vulnerable. Nobody gets to escape entirely, but given the variety of the effects, different areas and peoples are faced with different challenges in adapting to the effects of climate change. Different societies, classes, and peoples also contribute to climate change in different ways and degrees, be it through consuming electricity from coal-burning power stations, leading carbon-intensive lifestyles more generally, cutting down the forests that act as sinks and stores of carbon, or practicing forms of agriculture that lead to high emissions of greenhouse gases (including methane as well as carbon dioxide).

"Society" in our title also emphasizes the encompassing character of the climate challenge, which reaches into many if not most areas of human (and non-human) life. Climate change is actually just one aspect of global environmental change more generally; it can be thought of as just one "planetary boundary" (Rockström et al., 2009) that is being transgressed, one that interacts with other boundaries. (We will say much more about planetary boundaries in Chapter 7.) One of the other boundaries is ocean acidification, which is threatened by carbon dioxide (CO_2) absorbed into the ocean from the atmosphere. Some advocates of "geo-engineering" solutions (which we will also discuss in Chapter 7) suggest seeding the oceans with iron so they can absorb more carbon dioxide, which would of course make acidification worse. Climate change also has major implications for economic systems, lifestyles, government, development, social justice, poverty, transportation, health, security, and conflict. Moreover, what happens in all these other areas helps determine collective capacities to respond to climate change issues.

IMAGINE A RATIONAL WORLD

Imagine a Rational World Confronted by Climate Change. In this world, scientists would provide the basic description of the problem to be solved. They would, for example, ascertain the existing trajectory of greenhouse gas concentrations in the atmosphere and identify the consequences for the Earth's physical and ecological systems (such as rising sea levels, increasing frequency and severity of extreme weather, changing rainfall patterns, biodiversity depletion, ecosystem shifts, desertification, heat stress). Other experts (such as economists) could then help estimate all the future damages associated with different global temperature trajectories. This scientific and social knowledge would then be accepted by the world's publics and policy makers and inform a collective global agreement about overall targets. Perhaps global society would set the maximum increase in average global temperature compared to the pre-industrial level that should be allowed—as of 2009 that increase was 0.8°C. Alternatively, global society would decide the maximum allowable concentration of CO_2 (the most important, but not the only greenhouse gas) that should be allowed in the atmosphere—as this book was nearing completion, the level rose above 400 ppm, in comparison to the pre-industrial level of around 280 ppm. A timescale would then be put on the targets, and how the trajectory of future global emissions needs to look if such targets are to be met.

Let us say global society seeks stabilization at 450 ppm of CO_2. This level, according to Hare and Meinshausen (2006: 131), gives us a 53 percent chance of limiting the global average temperature increase to 2°C (the "guardrail" specified in the 2009 Copenhagen Accord reached at the end of a Conference of the Parties to the United Nations Framework Convention on Climate Change). We might get a trajectory that looks like that in Figure 1.1. This "contraction and convergence" graph has the radical implication of CO_2 emissions from fossil fuels falling to near zero by the late twenty-first century, implying a thoroughly de-carbonized global economy. The reason it has to look like this

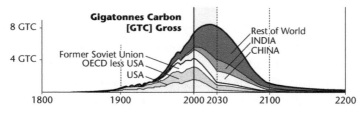

This example shows regionally negotiated rates of C&C.
It is for a 450ppmv Contraction Budget, with Convergence by 2030.

Figure 1.1. Contraction and convergence

Source: Global Commons Institute 2010, reprinted by permission.

is that once in the atmosphere, emissions accumulate. Now, half the historical emissions caused by humans have to date been absorbed by carbon sinks, on land and in the oceans. However, such sinks have a limited future capacity (Steffen, 2011: 26). Meinshausen et al. (2009) argue that taking the 2°C guardrail seriously allows total global emissions of 1000 gigatonnes of CO_2 in 2000–2050; yet in 2000–2009 alone 350 gigatonnes were emitted.

Once the trajectory of total global future emissions was agreed upon, a global treaty negotiated by all the world's countries would then allocate the required cuts and the timetable for achieving them to different countries (possibly even to all of them). One potential allocation is also represented in Figure 1.1, which has different categories of countries converging in their per capita emission levels by 2030, and all moving uniformly toward near zero thereafter. Each national government would then develop a set of policies to achieve these targets, taking advice from specialists such as economists as to what might work most effectively. Producers and consumers would comply with these policies, and adjust their behavior accordingly, until thorough de-carbonization of the economy is achieved, and emissions of other greenhouse gases are brought under control.

This rational imagery underlies current global efforts to respond to the climate challenge. The science is organized by the Intergovernmental Panel on Climate Change (IPCC), which produces periodic assessments that digest and summarize scientific findings and projections. Global agreement on overall targets and timetables as well as the allocation of greenhouse gas emissions reductions to particular countries is sought by the negotiations conducted under the auspices of the United Nations Framework Convention on Climate Change (UNFCCC) (established in 1992), most publicly in the annual two-week Conference of the Parties (COP). National governments are mostly committed to doing something in response.

9

WHAT HAPPENS INSTEAD

Unfortunately none of the elements of this smooth rational process comes anywhere close to being achieved in practice. To begin, though there is near-unanimity in the relevant scientific community about the broad realities and consequences of climate change, there remain some uncertainties concerning the precise links between the rate of growth of emissions, the concentration of greenhouse gases in the atmosphere, global average temperature, and the various resulting effects. These uncertainties underwrite different views of what the global target should be; for example, the activist group 350.org, as its name implies, seeks a world in which the level of CO_2 in the atmosphere is stabilized at 350 ppm. This is the same figure specified as a "planetary boundary" by Rockström et al. (2009)—which unfortunately has already been exceeded.

But these uncertainties pale into insignificance once the science enters the public arena, where (especially in the US) it proves controversial and contested; likewise the estimates of damages associated with different global temperature trajectories. The controversy in question is not of the kind that is normal in scientific communities as investigators test and challenge findings and hypotheses. Rather, it puts scientists in a pitched battle fought on very public terrain, with powerful political interests financing attacks on the integrity, content, and process of climate science—along with personal attacks on its practitioners. There are people who have much to lose if the overwhelming consensus among climate scientists is right: their lifestyles, their profits, and, in some cases, their livelihoods. If they look at Figure 1.1 they will see that the future holds a world with no place for coal and oil production, no place for fossil fuel companies, no place for energy-intensive lives built on fossil fuels.

In the US, Canada, and Australia, those who have had much to lose organized politically to challenge the scientific consensus (Oreskes and Conway, 2010). While the opponents of the near-unanimous scientific view can recruit a few dissident scientists, most of the latter are not actually climate scientists, and most of the battle is not fought among scientists, but rather between scientists and often skilled and ruthless political operators. Though they have political friends too (for example, former US Vice President Al Gore), climate scientists find they must enter the public arena. Often, however, they get very bruised when they do so (Schneider, 2009; Mann, 2012).

Within the economics of climate change, the situation is a little more complicated. In contrast to the broad consensus that characterizes climate science, there is legitimate contention among credible economists. Most of the disagreement centers around what may seem like a technicality: the appropriate rate of discount to apply to future benefits and costs. (We explain the discount rate concept in Chapter 3; essentially it is the way we value what happens in the future compared to what happens in the present.) However, this apparent technicality determines whether we treat the atmosphere as just another investment asset on the one hand, or as being in trust to us with obligations to future generations on the other. Disagreeing about the appropriate discount rate to apply, economists therefore also disagree with each other about the appropriate size and timing of global emissions reductions. Economists also disagree about how to handle risks and uncertainties.

If we move to the next step of our imaginary rational model, we find that a comprehensive global treaty on emission reduction is proving elusive. The Kyoto Protocol negotiated in 1997 committed developed industrial countries (so-called "Annex One" countries) to emissions reductions targets. But the targets applied only to a period expiring in 2012. The US renounced the Protocol (which it had never ratified) in 2001, President George W. Bush declaring that short-term US economic interests came first. In 2011, Canada tore up its ratification. Even if they were met, the Kyoto targets would have only minimally impeded the growth of total global greenhouse gas emissions. Since 1997, the UNFCCC negotiations have proven ever more divisive, as various countries and blocs of countries (such as the

European Union (EU), and G77 bloc of developing countries) fail to agree on who should bear the burden of emissions reduction. Questions of equity between rich and poor peoples divide both economists and national representatives, for answers to them determine who exactly should shoulder how much of the burden of emission reduction. Moreover, the experience of the Kyoto Protocol shows that even when a country does agree to a target and a timetable for reducing emissions, there is little guarantee that its government will actually seek to achieve the target; most of the Annex One countries failed to achieve their 2012 target.

Even if a government tries to meet its commitments, it may not succeed. The policies adopted may prove inadequate in practice and fail in their implementation. Policy analysts, interest organizations, and politicians disagree about the best strategy to adopt (options include emissions trading schemes, carbon taxes, direct regulation of emissions, the capture and storage of emissions, planned reconfiguration of the economy and its energy systems, restricting population and economic growth). Even if the national policy community can develop a course of action, producers and consumers will not necessarily comply with the policy decisions in question. So companies may fail to report or disguise their true emissions, and governments may lack the monitoring and enforcement capacity to do much about it.

In short, in light of the model of apparently rational action we sketched, it seems that everything that could possibly go wrong has gone wrong. The result is that the increasing urgency of calls for action is accompanied by diminishing likelihood of substantial response, especially at the crucial global level.

WHY THE DISSOLUTION OF REATIONALITY?

It is frequently observed that climate change presents a truly "diabolical" (Steffen, 2011) or "wicked" (Rittel and Webber, 1973) set of problems. The challenges include multiple dimensions, high stakes, complexity, consequences of decisions (and indecision) stretched out over long time periods, and intersection with most other areas of human and non-human life. In this kind of context, we see the science under siege, the economics contested, the politics divisive, the psychology of responses puzzling. So is it this wicked character that makes issues around climate change often seem so intractable?

We can approach an answer by noting that taken in isolation, many of the particular problems do not look so bad. If we isolate each particular problem, we often find we know quite a lot in terms of the consequences of inaction for social systems, and the repertoire of available responses (though putting that knowledge to use can still be difficult). What we know much less well is how to comprehend and respond to the entirety of the challenge, because the different bits can interact in unexpected, counterintuitive, or poorly understood ways. It is the interactions between the different aspects (and with other issues) that really cause difficulties. To take just one example: if the government of a country wants to reduce its CO_2 emissions it can take advice from economists on what kind of policy—for example, an emissions trading scheme of the sort we will discuss in Chapter 4—to introduce. On paper it can look very straightforward. But as soon as we get to the politics, we find affected economic interests will do their best to build exemptions and compensations into the scheme. Established polluters using old technologies may be in the best position to secure exemptions and subsidies, thus imposing the burden of emission abatement (and so competitive disadvantage) on newer and cleaner operators. The result may be to entrench high-carbon technologies. Here the interactions are between the economics, politics, and technological aspects of the issue. Matters become more complex still if the scheme in question incorporates emissions trading across national boundaries and allows companies to purchase offsets in other countries to meet their targets. If the offsets are in the form of growing trees, those previously relying on the forest for their livelihood may be displaced, raising major questions of social justice. Or offsets that exist on paper may not actually appear on the ground (or if they do, may be sold to multiple buyers or lost to forest fire).

Yet this "wickedness" and complexity still does not make climate change unique; other wicked and complex problems exist. Perhaps human brains did not evolve in an environment where they had to deal with problems of this scale, scope, and complexity (Jamieson, 2011: 48)—but, however, there are plenty of problems that have come along that are very different from life on the African Savannah where we evolved, and we haven't done too badly with some of them. While the psychology of response to climate change is interesting (and we will discuss its important role in Chapter 2), the faults do not lie in our cognitive selves but rather in our social systems.

Taking a moderately long view (at least in human terms), human social structures evolved during the past 10,000 years of the Holocene era defined by an unusually stable climate. So the idea of rapid climate change might seem like something novel. Yet if we look at matters a little more closely, we see that some particular human societies actually perished in the face of local climate change, though according to Jared Diamond (2005) there was nothing inevitable about the collapse, just a failure to adapt. If we examine social, political, and economic systems as they have developed in the past few centuries, they are highly adapted to deal with three kinds of problems: war, economy, and welfare. History does of course show periodic failure on all three fronts. But failure begets collective mobilization. When it comes to war, we see massive coordinated effort to reshape the international political system in the wake of total war through treaties in 1648 (Westphalia), 1815 (Vienna), 1919 (Versailles), and 1945 (Bretton Woods and then the United Nations). When it comes to economics, look at the staggering sums of money committed to saving the global financial system by the governments of wealthy states in response to the global financial crisis of 2008. When it comes to welfare, we see some developed countries redistributing almost a third of their national income through welfare states. The long-term trend has been in the direction of less war, continued economic growth, and, at least until recently, more welfare (in the form of income security).

Climate change is an altogether different kind of problem. Climate change seems to demand a degree of large-scale, collective, multi-faceted, coordinated, persistent, public-spirited, self-sacrificing, and—crucially—anticipatory response of a kind never really seen before in human affairs. It challenges the very character of a global civilization that has been built on fossil fuels. And as if that were not enough, it is driving home the need for humanity to negotiate a historical transition from the climatically stable Holocene that enabled agriculture and so complex human societies to develop, to the "Anthropocene," where it is the unintended consequences of human activity that increasingly shape the global environment in which we live.

Taking climate change seriously changes everything, from political systems that seem to require continued economic growth to secure their legitimacy and so survival, to cultures of mass consumption that everyone (especially those who currently don't have access) seems to want. We have already noted that society can occasionally generate massive coordinated response to collective problems—but that response is a reactive one in response to clear catastrophes that have already occurred on a very short timescale (for example, total war, or global financial crisis). Climate change is not like that; it is slow burning, and the effects of actions and omissions show up in places that can seem very distant in time and space.

RESPONDING TO THE FAILURE OF RATIONALITY

How then should we respond to this apparent wholesale collective failure of rationality? To date the response on the part of pretty much all the actors we have identified, from governments to international organizations, from scientists to activists, from lobbyists to consumers, has been to do more of what they have been doing for some time. Scientists conduct more research and produce more information, economists generate more cost estimates, governments continue to negotiate with each other in a multilateral forum, lobbyists and activists continue to try to influence those negotiations, climate change deniers and their financial backers continue to undermine science and action, enlightened consumers can continue to buy "green" goods. A definition of insanity often attributed (probably mistakenly) to Einstein is that it involves doing the same thing over and again and expecting to get a different result. In this light, the only actors who escape the charge are the organized climate-change deniers, because the results they are getting are largely what they want.

But it is not easy to identify what those who care about climate change and its impacts should do instead. Invoking catastrophic scenarios and lamenting the wicked, intractable character of the issues has its place, but may well have reached the limits of its utility. We know from studies of public opinion that scaring people no longer has much effect on motivation to do anything (Moser and Dilling, 2007). Action though still needs to be informed by understanding; and it is this kind of understanding that we hope to provide in this book. We do not claim to provide all the answers, but we do at least hope to confront all the perplexities head-on, identify the right questions, and do our best to respond to them. This may involve rethinking some entrenched understandings.

Responding to climate change is going to be a permanent feature of the human condition. If greenhouse gas emissions were to stop tomorrow, the continuing consequences of climate change would be played out for decades to come, average global temperature would continue to rise, and adaptation to a changing world would still be necessary. Of course emissions are not going to stop—indeed are likely to continue increasing for years to come (though at what rate and for how many years depends on a host of economic and political factors). Like it or not, ecological systems and the social systems that depend on them are going to be in a state of flux for any foreseeable future. Thus intelligent thinking must attend to adaptation to this changing world—while still keeping an eye on possibilities for reducing emissions and otherwise combating their effects in the aggregate.

HOW (ELSE) TO THINK

One response to the seeming failure of collective rationality in the face of climate change is to think of ways of redeeming the rational model we set out by overcoming the obstacles to it, one by one. Alternatively, we can try to think in different directions. Climate change is here to stay, a permanent consideration in many if not most realms of action. This does not mean giving up on a long-term goal such as de-carbonization of the global economy, still less downgrading the importance of the challenge, but rather recognizing climate change's omnipresent, pervasive, permanent, and encompassing character. We might even think about what climate change *enables* rather than *demands*. As Hulme (2009: 326) puts it, "the idea of climate change should be seen as an intellectual resource around which our collective and personal identities and projects can form and take shape." In this light, climate change enables thinking about community initiatives, not just rampant individualist consumerism. It enables thinking about how to reconstruct economies at any scale in ways more conducive to human values. It enables thinking about how to pursue justice in a world where the consequences of climate change are distributed so unevenly, and completely out of proportion to responsibility for the causes. It enables contemplation of just why our collective decision processes seem so maladapted to confront the issues that climate change presents, and how we might go about changing that. It enables confronting some profound questions about the character of our cultures. It enables rethinking the relationship between expert and lay knowledge, and how citizens might themselves participate in the generation and validation of knowledge.

But this switch from "demands" to "enables" has to be more than linguistic trickery (try changing "enables" to "demands" in the previous six sentences; they still make sense). Effective response to climate change requires nothing less than solving some of the persistent riddles of human environmental history. How can a recalcitrant social world change so as to be more sustainable? If not through choosing better policies, is it a matter of structural change in the core institutions of the political economy? Or of shifts in the terms of discourse about not just environmental affairs, but everything that affects them? Or cultural change? We suspect it is all of these things: framing issues in terms of opportunities and more felicitous relationships with pursuit of a range of social goods (such as conviviality, security, justice) rather than threats may help along the way, but by itself does not unlock the door to a happier future.

Here we need to steer a course between a totalizing response that seems not to deliver (the rational model we set out) and "anything goes"—a useless Sinatra doctrine for the environment. (The original Sinatra doctrine was proclaimed by former Soviet President Gorbachev, so that each country in Eastern Europe could go "My Way.") Responding to climate change requires many things, not one big thing. But these should not be thought of as small, disconnected things. They need to connect in order to constitute a response that is at once profound but hard to specify in fixed, final form. Re-stating the challenge in this form drives home its novelty. The very need for pluralism plus connection suggests a different kind of world: one that is not subordinated to hierarchy, conflict, or competition, where no single body deserves to be the repository of all our hopes, but rather where networks and creative, cooperative relationships play key roles. Of course networks themselves come in many varieties; for example existing governance networks are often low-visibility and dominated by established powerful interests (see Chapter 6). So invoking networks is not of itself the answer;

it is just indicative of the need to think through some basic questions about how human societies operate and are organized, and how multiple human endeavors can be linked in ways that are different to existing inadequate social, economic, and political systems. Now let us start.

2

Constructing Science and Dealing with Denial

PRODUCING CLIMATE SCIENCE

Climate science has a long history. The Swede Svante Arrhenius in 1896 recognized that the burning of fossil fuels could add CO_2 to the atmosphere in sufficient quantities to warm the Earth, though he thought it would take millennia for that to become apparent. Arrhenius himself thought this would be beneficial to agriculture, anticipating some contemporary emphatic climate change deniers for whom CO_2 is nothing more or less than "plant food." The twentieth century saw anthropogenic (i.e. caused by humans) climate change gradually progress from a scientific curiosity likely to arise only in a very distant future to something more pressing (see Weart, 2008 for a history). Charles Keeling began monitoring atmospheric CO_2 on Mount Mauna Loa in the middle of the Pacific Ocean in 1958, providing strong evidence that CO_2 levels were rising. In 1965 the Science Advisory Committee to the US president raised the specter of changes in the climate appearing by 2000. Climate science gradually grew in extent and prominence, aided by advances in satellite monitoring and computing power.

One watershed moment occurred in 1988, on a hot day in Washington DC, when James Hansen of NASA testified to the

Energy and Natural Resources Committee of the US Senate that global warming had arrived. The same year British Prime Minister Margaret Thatcher (who had a degree in chemistry) announced in a speech to the scientists of the Royal Society that she was convinced of the need to act—embracing environmental concern she had until then derided.

Since the 1980s climate research has exploded, exploring ever more facets of the issue. The role of the IPCC, established by the United Nations in 1988, has become crucial. The Panel does not actually conduct or sponsor research itself, but rather summarizes the weight of scientific opinion in periodic assessment reports aimed at policy makers, especially those participating in the negotiations of the UNFCCC. With literally thousands of scientists from diverse disciplines participating in the assessment, it has a significant impact on how scientists connect their subsequent research to discoveries by others and learn how to communicate with each other, building an ever greater capacity to both assess and synthesize climate science into a more cohesive whole (Edwards, 2010). At this writing, the IPCC is engaged in its fifth assessment effort to be completed in 2014.

THE STATE OF THE SCIENCE

What then is the state of the science? There is broad agreement on at least seven propositions and their implications for the nature of the challenge of climate change:

1. The measurable acceleration in climate change over the last few decades is extremely likely to be a result of ever-increasing greenhouse gas emissions stemming from industrial development, growth in transportation, global deforestation, and other changes in land use.

2. Many of the factors contributing to climate change are cumulative and long-lived. Even if greenhouse gas

emissions were to fall to zero today, and if forests and other carbon sinks stabilized, it would take several centuries for the climate system to return to its pre-industrial condition through the slow absorption of carbon in the ocean and its eventual sequestration in the shells of marine animals. This means in turn that benefits from any current reduction in greenhouse gas emissions will be distributed over the next several centuries.

3. The effects of climate change are not distributed uniformly. For example, warming near the poles is generally greater than toward the equator.

4. Climate models indicate that as average global temperatures increase, coastal areas are likely to be wetter while continental interiors are likely to be hotter and drier, though there are substantial local variations in these effects.

5. Average global temperature increase is accompanied by greater variation in weather. Extreme events such as hurricanes, heat waves, and droughts become more frequent (though any particular event cannot be attributed with any confidence to climate change).

6. Sea levels that were relatively stable for several centuries prior to 1900 rose at about 1.5 millimeters per year on average during the twentieth century. Levels now appear to be rising at twice that rate, and during the twenty-first century are expected to rise between 0.7 and 1.4 meters in total. The ocean currents that distribute heat from equatorial regions toward the poles and pile up water in some places more than others are affected by climate change. Higher ocean temperatures affect levels of precipitation and intensity of storms, and intensity of storms and winds can affect sea levels.

7. Much of the uncertainty about climate dynamics stems from the pretty much irreducible uncertainties concerning how forests and soils, the most significant stocks of biologically sequestered carbon, can both amplify and moderate physical processes in the climate system.

How quickly the climate will warm for any given path of global greenhouse gas emissions remains uncertain, as does the precise nature of a potentially catastrophic changed planetary state to which unchecked climate change might eventually lead. Williams et al. (2007) suggest that by 2100, the climates we currently see on 10–48 percent of the Earth's surface may have gone, and entirely novel kinds of climate may exist on 12–39 percent of the planet (though they allow that other scenarios produce lower estimates). Uncertainty also remains concerning how various effects will play out around the globe—rates of sea level rise, precipitation changes, the frequency of storms and wildfires, the incidence of species loss. Scientists have been surprised that some effects, such as melting of sea ice in the Arctic, have arisen as early as they have.

Even as these details remain uncertain, climate scientists generally assert that there is a scientific consensus on the reality and significance of climate change, and this is true enough. Why then can critics impede public validation of this consensus, at least in Anglo-American countries? The answer may be that there are enough uncertainties and variations in the content of the science to provide ammunition for the critics (who have no coherent body of science of their own to deploy). In light of the politically charged environment in which they operate, it is perhaps no surprise that climate scientists themselves stress consensus in their public pronouncements. Yet if we probe a little deeper, we do see a measure of disagreement, for example when it comes to the likely future trajectory of average global temperature associated with particular trends in greenhouse gas concentration in the atmosphere, or what series of particular indicators actually show, or how to reconcile different sorts of data that exist for different time periods.

A degree of disagreement is of course normal in any science, but the nature of the subject matter of climate science provides additional possibilities for dispute. Climate science makes some physicists a bit nervous, for there is no possibility for controlled experiments of the sort central to the testing of hypotheses in many other branches of physics (von Storch et al., 2011: 115–16). Moreover, some of the big questions concern what is likely to happen in a future that has not yet arrived, taking the form not of exact predictions, but rather conditional scenarios and projections (von Storch et al., 2011: 119). The possible existence of tipping points (where a small change in one factor may induce feedbacks sufficient to cause large changes in the system as a whole) introduces further uncertainty into projections. Barnosky et al. (2012) suggest there may be tipping points at the planetary scale that could even mean a "state shift" for the biosphere as a whole if current human-induced changes (including climate change) continue. Such shifts have occurred in the Earth's past, sometimes accompanied by mass extinctions. Obviously any such shift would take the world into unknown territory.

Climate science deals with complex systems. And one of the hallmarks of a complex system is that it is only amenable to analysis using multiple and different frameworks and strategies for analysis. So, for example, in ascertaining the significance of current warming trends, it is necessary to construct a time series of average global temperature. This can be done using data from a geographically dispersed set of monitoring stations. Yet this would not take us back beyond the middle of the nineteenth century, so for a deeper history scientists can look at evidence from tree rings or cores drilled into the ice of Antarctica or Greenland. To get a complete record of the sort that appears in the famous "hockey stick" graph (see Figure 2.1), disparate data from different sources must be combined.

Different methods for making unrelated raw data comparable do not give exactly the same results. Thus there exists a range of uncertainty concerning the exact historical trajectory of global temperatures; and still more concerning what that past trajectory tells us about likely futures. This opens up the opportunity for partisans of particular political or ethical

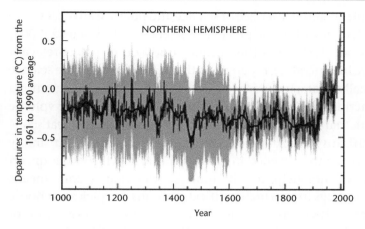

Figure 2.1. The "hockey stick" representation of global temperature over time
Source: Intergovernmental Panel on Climate Change, Third Assessment Report, 2001. Recent data from thermometers and earlier from tree rings, corals, ice cores and historical records. The original "hockey stick" was published in Mann et al. (1998).

positions to find information that supports their own position most strongly. So for example climate change deniers stress the existence of a medieval warm period that was clearly not caused by greenhouse gas emissions; or global temperature trends for the decade following 2000 that actually shows cooling because 2000 was an exceptionally hot year. Advocates of international climate justice in contrast could highlight the fact that warming trends are unlikely to be uniform across the globe. In particular, sub-Saharan Africa is likely to experience warming 50 percent higher than the global average (so, if the world warms by 2°C, sub-Saharan Africa warms by 3°C), thus leading to especially damaging impacts in that already impoverished region. Of course there is much more to climate science than global temperature trends and regional variations. The likelihood of changes in the patterns and timing of rainfall are also critical. Thus there is plenty of opportunity for partisans of any sort to be selective in the bits of evidence they highlight.

Problems at the interface of science and political advocacy cannot be cured within the practice of science itself. Sarewitz (2004) points out that this is the normal condition of policy-relevant science in general. Compared to other such areas, climate science is actually characterized by a low degree of disagreement among its practitioners; it stands out only in terms of the sheer volume of work that gets done, thus producing exceptionally great scope for partisans to pick and choose findings. It follows that simply producing more science in the hope that it will reduce uncertainty will not help, for the more science that is done means more findings from which political partisans can pick and choose to support their case. Science itself, as Sarewitz stresses, cannot resolve disputes that are at root political and/or ethical.

THE COMMUNICATION AND RECEPTION OF SCIENCE

The subtleties of irreducible uncertainty tend to get lost in the communication of climate science to the public. Perhaps the most well-known communicator is former US Vice President Al Gore, awarded a Nobel Peace Prize in 2007 for his efforts including a global lecture tour and his 2006 Oscar-winning film *An Inconvenient Truth*. This documentary was designed to convince those who viewed it of the magnitude, severity, and immediacy of the risks associated with anthropogenic climate change. The film was a popular success in many countries.

Gore's efforts are, as Moser and Dilling (2011: 163) point out, consistent with an "information deficit model" of science communication: if only members of the public knew enough they would act. But this model turns out to be woefully incomplete: more knowledge does not necessarily lead to behavioral change or political action. Scaring people with doom-laden scenarios about the effects of climate change also proves counterproductive (Moser and Dilling, 2011: 164–5). Insisting on the authority of science does not always help. For climate science, especially when it identifies global changes, challenges and disrupts some of the most cherished social, economic, and psychological investments that particular societies have made in their own futures and the commonsense understandings that go along with these investments (Jasanoff, 2010). These investments include national self-determination, economic growth, and freedom to consume. Just insisting that science should win this confrontation will not do.

Kari Norgaard (2011) identifies a particularly subtle problem in the reception of climate change communication. Norgaard looks at Norway, instructive because of its high level of policy commitment and public concern on environmental issues. Yet Norgaard finds that all kinds of psychological mechanisms enter to prevent people from fully accepting and acting upon their knowledge and their value commitments. People are quite capable of accepting the reality of damaging anthropogenic climate change on the one hand, but not letting that knowledge have any influence on their lives on the other hand. So for example people can recognize the existence of the problem but not their own share of any responsibility for its cause or solution, can think of it as something to be solved by others. Norwegians can even tell themselves that their development of oil and gas resources for export is actually beneficial for the global climate— because the alternative sources of fuels (especially coal) are much less efficient in their extraction and production and much dirtier in their effects. If these sorts of psychological mechanisms operate in Norway, of all places, then they can probably operate anywhere.

Acceptance of and resistance to climate science can also be understood in light of existing beliefs, ideologies, and identities. Whatever it is that climate scientists say, its reception is filtered by ideology. This is not simply an "irrational" response to the science; it is a way to think about that information in light of one's larger system of beliefs. Human beings seek a sense of consistency in their positions—a single story into which beliefs can fall. Once they make up their minds, more science or evidence will not change their position.

This filtering process can be understood in at least two ways. On the one hand, interpretations of climate science are conditioned by political partisanship. In a polarized political system like the US, one of the strongest predictors of whether a citizen agrees with climate science is their political party identification (McCright and Dunlap, 2011). A change in position of the party seems to change minds more than additional science; we see a drop of levels of belief in climate change among conservatives as the Republican Party in the US (and the conservative Liberal-National Coalition in Australia) became more hardline in their denial.

Dan Kahan and his colleagues see a deeper process of "cultural cognition" at work where people with basic differences in their cultural values—hierarchical individualists on the one hand versus egalitarian communitarians on the other—draw different conclusions from the same information (Kahan et al., 2010). At least in the US (though not necessarily elsewhere), egalitarian communitarians just believe in the climate science more than individualists do. Communitarians are more concerned with the social good, and so we would expect this cultural cognition to lead to more sympathy for not only the science of climate change, but also the collective need to respond.

Reception of climate science and its implications can also be influenced by religion. Religion can motivate morality to protect future generations, those most vulnerable to the effects of climate change, and nature (Gottlieb, 2006). Many people of faith accept the science and the evidence and are climate activists—in, for example, the "creation care" movement among evangelical Christians (Kearns, 2011). There are, however, many Christians, especially in the US, whose religious beliefs come into conflict with what they see in the science.

Religion, like science, is one way of satisfying what seems to be an innate human desire to understand the world, to feel there is underlying consistency and coherency in the complexities of nature and society. For most of recorded human history, religious explanations filled this need. In modern times, science promised it could do a better job. In the words of biologist Edward O. Wilson (1998: 297): "The legacy of the Enlightenment is the belief that entirely on our own we can know, and in knowing, understand, and in understanding, choose wisely." Also, "When we have unified enough certain knowledge, we will understand who we are and why we are here" (pp. 6–7).

Science's challenge to religion was ameliorated for centuries through a division of labor: science would develop a consistent and coherent explanation of the world around us, while religion would deal with ethics and meaning. Yet some people still found religious explanations of reality more complete and comforting. And with time, scientific explanations of how things fit together became increasingly fragmented, complex, filled with uncertainties, and, crucially, beyond the grasp of individuals, even highly trained individual scientists. While the proportion of people relying on religious explanations had seemed to decline as science progressed, the way science eventually developed perhaps renewed the appeal of religious explanations.

Climate science both exemplifies this trajectory and undermines the idea that science can help us control nature for human benefit. Climate scientists now say that humans have far less control than once thought, and indeed created the circumstances through which they lost control, and became more subject to the whims of nature. If God's plan of human dominion has not changed then climate scientists must be frauds. In the US especially, fossil fuel interests have helped organize people with strong religious beliefs to oppose climate science by exploiting what looks like the failed promise of science, as well as its revealed challenge to God's plan (Oreskes and Conway, 2010; Kearns, 2011).

Even if all these obstacles can be overcome and a concerned public can be mobilized, getting the science accepted and acted upon by policy makers is still difficult. The way the science is processed in policy making turns out to vary substantially in different countries with different cultural contexts in which science gets interpreted and acted upon. Jasanoff (2011) describes a US setting that is adversarial—scientists can testify in support of different policy positions—where closure is sought through legal means. One example of closure is the *Massachusetts v Environmental Protection Agency* Supreme Court decision that in 2007 determined that greenhouse gases were pollutants and required the Environmental Protection Agency to explain why it was not regulating them. In Britain, in contrast, science gets validated through consensus on the part of relevant non-specialist policy elites prepared to take scientific findings on trust if there is confidence in the integrity of scientists. Germany seeks a more encompassing social consensus in the context of strong public aversion to risks. Among these three cases, it is the US where opponents of climate science have the most varied opportunities to press their case. The reason is that while there are moments of apparent closure, the sheer number of points at which policy proposals can be vetoed in the American political system means that partisans who want to prevent action always have the opportunity to find another step in the process (be it in a court, legislature, or executive agency).

SKEPTICISM AND DENIAL

The term "skeptic" is often used to characterize those who do not accept the existence of significant and damaging anthropogenic climate change. But this is actually a bit of a misnomer, because a skeptic in the true sense of the word is simply someone who does not take established wisdom on trust, who prizes data and reasoning. Skepticism in this sense is a valuable scientific attribute—in climate science no less than elsewhere. So the self-styled "skeptical environmentalist" Bjørn Lomborg, who made his name by questioning the warnings of environmentalists about resource scarcities and ecological degradation, can also say "it is vital to emphasize the consensus on the most important scientific questions. We have long moved on from any mainstream disagreements about the science of climate change" (Lomborg, 2010: 1). The hardline deniers of climate change are anything but skeptical in this sense, because there is no amount of climate science or argument based upon it that could induce them to change their minds. What McCright and Dunlap (2011) call "organized denial," financed by fossil fuel companies such as ExxonMobil and run by right-wing think tanks in the US such as the Heritage Foundation, aims at winning, not understanding or reasoning. Its task is to undermine and discredit climate science, not contribute to scientific debate, still less generate a body of science of its own (Oreskes and Conway, 2010).

Organized denial of this sort has its origins in the US but is also influential in Canada and Australia—not coincidentally, all three are Anglo-American settler societies with a frontier developmental ethos. This movement has been very successful in getting its message across in the media, and its partisans are particularly zealous when it comes to use of the internet to comment on any story pertaining to climate change. It has made inroads into political parties, notably the Republican Party in the US, where denial of the existence of anthropogenic climate change became necessary by 2012 for aspiring presidential candidates. The Republican Party certainly changed comprehensively since the early 1970s when President Richard Nixon introduced the most comprehensive suite of science-based environmental laws and policies in the history of the US.

In these three countries, organized denial has been successful in inducing doubt in public opinion, and legitimating the idea that there are two sides to the story when it comes to the existence of climate change. In the US, the "balance" doctrine in the mainstream media means that an effort is made to present both sides (but no more than two sides) on any issue that looks controversial: climate change is regarded as such an issue, thus institutionalizing a platform for denial (Boykoff and Boykoff, 2004). There are also popular media outlets in the US that make no pretense of balance or objectivity in their attempts to advance denial, such as Fox News.

Organized denial received a major boost with the publication of stolen emails from the University of East Anglia's Climate Research Unit in 2009, just before the COP of the UNFCCC in Copenhagen. Those emails contained some intemperate comments that looked like violations of scientific objectivity, and some loose language about using a "trick" in presenting data. There was actually little or nothing in those emails that would not in all likelihood be found in any similar volume of communication in any scholarly community. An inquiry carried out by the Select Committee on Science and Technology of the UK House of Commons subsequently cleared the scientists involved of any scientific wrongdoing. But the damage was done; public trust was undermined. And trust in the messenger is crucial when it comes to getting public acceptance of specialist knowledge of any kind. Climate scientists came to feel increasingly beleaguered, subject to all kinds of frivolous requests for information about their research, and many very personal attacks.

In most countries climate change denial is a minor political force. But is there any way to engage those who do not accept the reality of anthropogenic climate change in productive debate in the US, Canada, and Australia? When it comes to the organized denial campaign, the answer is surely no. But not all of those who do not accept the consensus of climate science fall into this camp. Some are indeed skeptics in the true sense of the word. So let us take a close look at the varieties of denial and skepticism.

REACHING SKEPTICS

Here it might be possible to begin with a categorization of skeptics and deniers. The different types might be:

- Those who deny that climate change could ever exist.
- Those who believe it could exist, but is not caused by humans.

- Those who believe it exists, but does not do any substantial damage.
- Those who believe it exists, is damaging, but nothing can be done about it.
- Those who think that something should be done—but not by themselves.
- Those who are unsure about what is happening and require convincing.

The reality is a bit messier: people do not fall neatly into these types. Instead, as Hobson and Niemeyer (2013) find, particular individuals mix these sorts of positions. Hobson and Niemeyer proceed to engage in a more inductive kind of analysis of skeptics in two Australian communities, which reveals a variety of positions ranging from what they call "emphatic negation" through to the "noncommittal consent" of a group of people genuinely uncertain about what is happening.

Even in the US, Canada, and Australia, the hotbeds of skepticism and denial, this last category is probably much larger than hardline emphatic denial. It is hard to say just how large this group actually is. There exists a considerable amount of opinion polling on climate change in different countries. But results obtained vary with the way the question is framed, and what cues are provided by the survey. Depending on how (and where) the question is asked, up to 50 percent of respondents may say that the case for anthropogenic climate change has not been proven. Logically, even if one does not think dangerous climate change has been proven, one might still think the risk sufficiently high to warrant policy action (when it comes to the tiny risk of being the victim of a terrorist attack, popular majorities can support massive and expensive preventive action). But if climate change does indeed present a challenge of the magnitude described in Chapter 1, it still seems there are many people who remain to be convinced about the need for action, and in a democratic society, this matters. How might climate change be communicated to them?

Here it is easy to point to what no longer seems to make much of a difference (more information, scaring people, asserting the authority of the science), hard to point to what can be more effective. We might take advice from public relations professionals about how to select the right messenger (for example, one who can generate trust, and not polarize the audience) and how to make messages palatable; from decision psychologists about the relative efficacy of scaring people versus accentuating the positive in terms of opportunities for individual and collective action; from advertisers about the kind of messages that hit home; from students of rhetoric about the ways people can be persuaded. Yet these strategies still rely on the idea of a mass media that reaches people and can lead them to see the world differently and act upon that knowledge. This very assumption may be at fault; Moser and Dilling (2011: 168) survey the evidence from different sorts of issues (notably health) and conclude that media campaigns are far less effective than face-to-face communication.

SCIENCE IN SOCIETY

Communicating climate change better may prove impossible so long as we assume that climate science and climate change are something to be communicated from experts *to* a lay audience, or a general public. Let us now take a look at the idea that the relevant knowledge should be produced *with* people outside the scientific community.

In one sense this development is inescapable: climate science is already wrapped up in politics, as scientists find they must make their case in a political process, have their research funding affected by political priorities, must deal with the throng of organized deniers trying to waylay them at every turn, and see that their results are deployed in ways that support or undermine particular partisan agendas. Stepping back from the fray in the name of scientific aloofness and political neutrality is no longer an option.

Rather than bemoan this engagement, it could be accepted and utilized. Moser and Dilling argue that a corollary of the failures they have noted in one-way mass communication is the need for more dialogical face-to-face engagement encompassing experts, advocates, and publics, where trust can be built, reflection can occur more readily, and experts can respond directly to lay concerns. The science *per se* would of course continue to be the preserve of scientists. But publics can enter in setting the agenda for the questions that science needs to answer, deciding what problems merit attention, interpreting the significance of findings, integrating scientific knowledge with lay concerns, and recommending policies. This kind of process already occurs on some issues in some jurisdictions, under headings such as participatory technology assessment. Several projects have been undertaken on climate change; for example, the Alberta Climate Dialogue organized in 2010–2015, involving extensive face-to-face citizen deliberation on the province's climate change policy, including the controversial (highly inefficient in terms of net energy produced) extraction of oil from tar sands. Such projects normally involve a forum in which citizens are exposed to the complexities of an issue, get a chance to question experts and advocates, deliberate with each other, and reach considered judgments.

There is some evidence that involving climate skeptics in such forums can have salutary effects (Hobson and Niemeyer, 2013). Some might become more convinced of the need for collective action on climate issues. Surprisingly, even those who remain unconvinced of the evidence for anthropogenic climate change may still be willing to enter productive discussion about policy alternatives such as a carbon tax (Lo, 2011). While according to a conventional model of individual rationality this seems illogical, what it demonstrates is perhaps the variety of motivations that underlie apparent skepticism—in this particular case, mistrust of government, which could be ameliorated by careful attention to policy design, so that a carbon tax was not seen as just a cover for government expansion. The evidence suggests though that deliberation does not affect the positions of hard-liners: Hobson and Niemeyer show that individuals whose initial position is "emphatic denial" actually hardened their views after participating in a citizen forum that involved testimony by experts, lots of locally specific information about the impacts of climate change, and deliberation with fellow citizens.

This kind of participatory approach does though face its own challenges. It is very hard to involve more than a tiny proportion of citizens in face-to-face public engagement processes. It might be possible to embed such forums in larger public processes involving old and new media and more traditional sorts of political action. But when Australian Prime Minister Julia Gillard proposed a citizens' assembly to deliberate climate policy in the context of a general action campaign in 2011, the proposal was widely ridiculed as a move designed to cover her government's inability to craft effective policy—not just by a conservative opposition party containing climate skeptics and deniers, but also by the Green Party and the country's leading allegedly progressive think tank (the Australia Institute), contemptuous of the idea that ordinary citizens might be able to make a constructive contribution.

What this Australian case really illustrates is the difficulty that political systems used to treating every issue as grist for a two-sided partisan electoral contest have when it comes to processing knowledge on an issue like climate change. Proposals for deliberative forums are not necessarily introduced in such inauspicious circumstances in such inhospitable political systems. It is possible to imagine much more productive participatory and deliberative engagement joining policy makers, scientists, and citizens—though the challenge here is not just one of the relationship between science and society, but also one of how we organize governance. We will have much more to say about the challenge of governance and how more productive forms—deliberative forms in particular—can help respond to this challenge in Chapters 6 and 8.

CONCLUSION

In environmental controversies, it is normal to see conflict between competing values and interests: environmentalists versus developers, conservation versus economic growth, victims of environmental damage versus perpetrators, extractive resource users versus preservationists. Climate change is different in that much of the controversy surrounds the content of science. Perhaps it is its traditional claim to neutral authority that makes climate science look so dangerous to those whose values, interests, identities, and lifestyles are challenged by the very idea of damaging anthropogenic climate change. With the resulting attacks, science loses its monopoly of authority over questions of knowledge, and becomes politicized, whether scientists themselves like it or not. Contesting the science of climate change is one line of defense for people who think they have a lot to lose if society takes climate change seriously. Another line of defense against thoroughgoing action might conceivably be found in the economics: that is, the value that society places on the damage that climate change is likely to cause versus what it will cost to try to prevent that damage. We turn to the economics in the next chapter.

3

The Costs of Inaction and the Limits of Economics

The costs of climate change, like the proverbial "death and taxes," are inevitable, though not entirely fixed or predictable in terms of when they arrive. Humanity has some control over the specifics. As with taxes, different people will be suffering different levels—though when it comes to climate change, the damage can fall most heavily on those least able to bear it. In addition, the costs of inaction will mostly be borne by today's young people and their children and grandchildren. Thus moral issues arise concerning how the burdens of action should be shared. Many climate scientists see the costs of inaction as very likely immense, making inaction a foolhardy gamble that must be avoided. Many economists, by contrast, are still arguing over how to compute the net benefits of doing anything versus doing nothing.

We surveyed the dimensions of the likely damage due to climate change as identified by scientists in Chapter 2. In this chapter our concern is not with the actual content of the damage, but rather with how to put a value on it, and what this implies for the character and magnitude of actions that should be taken. When the costs of inaction are clearly greater than the costs of action, basic economic logic would seem to dictate that action should be taken.

Humanity will inevitably bear some mix of the costs of mitigating climate change (especially by reducing emissions), adapting to change, and living with consequences that are not avoided. Doing nothing to mitigate, as has largely been the case so far, results in the costs of inaction we will emphasize in this chapter. If inaction continues, the science tells us that the risks are huge—eventually the future of humanity and all of life as we know it are at stake. No economists advocate driving humanity to ruin, but many seem willing to gamble with that possibility in exchange for the benefits of faster economic development through continuing exploitation of fossil fuels in both the short and long term. We take a close look at what economists can usefully contribute in a climate-challenged society, but also show what still needs to be done even after economics is stretched to its limit in taking climate change seriously.

THE OZONE PRECEDENT

There has been only one global environmental issue where an economic calculation based on scientific understanding seems to have played a key role in driving decisive public action. In 1987 the world's nations, led by the US, agreed to the Montreal Protocol on Substances that Deplete the Ozone Layer. This agreement restricted the use of chlorofluorocarbons (CFCs) used mostly in refrigerators and air conditioners and as a propellant in spray cans. An unusual combination of factors contributed to speedy agreement once it was clear that CFCs were indeed impeding how harmful ultraviolet radiation was filtered out in the stratosphere. The costs of inaction seemed unbearably high. For the US alone, nearly a million premature deaths due to cancer were estimated to be avoidable between 1986 and 2075 by reducing global CFC use by only 20 percent. The US Environmental Protection Agency, backed by the Council of Economic Advisors, estimated the monetary value of these lives lost at $1.3 trillion, with the costs of reducing CFC use by 20 percent (the US share of CFC use) at merely $4 billion (DeCanio, 2003). The health effects of ozone depletion were far greater in other parts of the globe, especially Argentina, Chile, and New Zealand. With the costs of doing nothing exceeding the costs of action by more than 30 to 1 in the US alone, action was clearly appropriate.

However, even this massive disparity and very clear conclusion was not enough to yield global agreement, which was initially blocked by the EU, and only weakly sought by the US. A number of factors changed around 1987 to make agreement possible. The first was the rhetorical force of the idea of an "ozone hole" in the Southern hemisphere leading to a global shift in favor of what Litfin (1994) calls a "precautionary" discourse. The second was a subtle reframing of the EU's perception of its own interests that led it to change its negotiating position (Berejikian, 2004). The third was the fact that DuPont, the major US producer of CFCs, had developed and patented effective substitutes. DuPont, and so the US negotiators, could therefore argue that CFCs were not a technological necessity, as European and Japanese chemical companies had claimed, and that the economic costs of quickly reducing their use would be acceptable. DuPont, of course, benefitted through a gain of market share for its substitutes (Benedick, 1991).

The moral of the ozone story is therefore that even an obvious massive net benefit will not by itself guarantee global action. Yet by the late 1980s enough seemed to be known about the environmental implications of climate change that scientists expected, in retrospect a bit naively, a Montreal Protocol-like response. Surely, rather than face these costs, the planet's people would wean themselves from fossil fuels for the sake of the long-term well-being of humanity and indeed all life as we know it.

Such, however, did not turn out to be the case, initially in large part because the costs of doing nothing, at least as understood by economists, were not as obvious as for ozone layer depletion. In addition, the understandings of scientists and economists sometimes collided with the hopes and expectations of different publics, economic interests, and policy makers concerning how science and economics are supposed to work and inform policy. In Chapter 2 we looked at some of the strange things that can happen when climate science encounters powerful vested interests and confronts established ways of life. In this chapter we look at how economics attempts to clarify matters and guide what should be done—but also often ends up further muddying the waters. We will conclude that climate economics as conventionally practiced should not be accepted as an authoritative guide to action. This judgment applies to economists who attempt to grapple with the ethical dimensions of costs imposed upon vulnerable others (most famously, in the UK's 2007 Stern Review) almost as much as to an earlier generation of climate economists who downplayed these ethical aspects, to whom we turn first.

THE EARLY RESPONSE OF ECONOMISTS

In contemporary growth-oriented capitalist economies, economists play a special role as experts trained to address, "objectively," questions of public policy choice; their judgments on climate change have proven influential (though in practice, economic calculation alone rarely drives public policy, on climate change or on other issues). In 1990, William Nordhaus, already the leading climate economist, wrote in *The Economist* that global warming had both benefits and costs and that the balance of these against the costs of preventing warming was by no means clear. He argued that, while there may be a few very low-cost preventive measures that could be taken, anything more should wait for cool-headed research (Nordhaus, 1990). Two years later, Thomas Schelling chose to speak about climate change in his Presidential Address to the American Economics Association. Schelling pointed out that people in the US, especially retirees free to move, and economic activity generally had long been migrating to warmer parts of the country. Furthermore, more people died in cold spells than in heat spells, so a little warming would surely be beneficial (Schelling, 1992). Thus, at a time when most climate scientists were arguing for transition away from a fossil fuel economy, key economists were arguing the opposite. Defending this status quo also meant that economists were supporting established economic interests, and hence the current power structure, though that is not how they thought of themselves.

Climate economists eventually shifted toward advocacy of increasingly significant action to combat climate change. Nevertheless, fossil fuel interests and others with a stake in the status quo have kept economists' early arguments against doing anything alive, and continue to amplify them to the public, especially in the US (Oreskes and Conway, 2010).

Two decades of cool-headed economic research since Nordhaus made his plea for studied inaction mean that much more is now known about the dynamics of climate change and the costs of inaction, which prove considerably greater than they seemed to be in the 1990s. Yet the way most climate economists view the problem is still not in line with how climate scientists see it. Much of the difference is rooted in the way economists were originally brought into the policy process rather than in the nature of economic theory, so let us take a look at how policy-associated economics came to its current condition.

ECONOMICS AND PUBLIC POLICY

In the US, economists were brought into government during the Great Depression in the 1930s to help design and select projects to increase employment. After World War II, the advice of economists was thought to be essential to assess public investments in all industrial economies. Economists developed cost–benefit analysis to fill this public role (Chakravarty, 1987), which in turn created further opportunities for them to advise on public investment decisions (Nelson, 1987). This history is critically important in understanding the contemporary role of economists because the analytical precedents established actually prove inappropriate for big questions concerning climate change—though this has not stopped cost–benefit economists playing a major role.

At one level cost–benefit analysis is very simple: before deciding to do something, we need to enumerate all the costs and benefits of the proposed action, then see if the benefits outweigh the costs. More technically, the criterion of cost–benefit analysis is to maximize net present value (NPV), which involves calculating a present-day equivalent of all the costs and all the benefits that will be felt in future time periods, as well as the present one. So:

$$NPV = \sum_{t=0}^{n} \frac{B_t - C_t}{(1+r)^t}$$

where: t is the number of years from the present (so $t = 0$ is this year), Σ refers to the sum across years between the present and n years from now, B_t = benefits in year t, C_t = costs in year t, r = the rate of interest. Future benefits and costs are divided by $(1 + r)^t$, a procedure known as "discounting" to the present. The rationale is that $1 put in a savings account today equals $1.05 a year from now (that is, when $t = 1$) when the interest rate is 5 percent, and so the value of a benefit or cost of $1.05 in a year's time is equivalent to $1 today.

Many early applications of cost-benefit analysis in the US were to water projects. Cost–benefit analysis evolved in the context of spreading development from the industrialized east to poor regions such as the Tennessee Valley, delivering electricity to rural areas, and bringing water to the arid southwest. Questions surrounding investment in dams raised three broad issues. First, water projects entail major construction costs in the near term with benefits that stretch out over long time periods, meaning choice of discount rate for public investments is critical.

The lower the discount rate chosen, the higher the long-term benefits look when they are converted to the present. Second, estimating the benefits of water development, and later its environmental costs, required new valuation techniques. There is no market price for large recreational lakes created by dams or for the fishing streams lost, so economists had to develop indirect measures (for example, calculating the value of a fishing stream by estimating how much people pay in travel and gear to fish in it). Third, benefits and costs are not evenly distributed between people, and when some people seem to benefit while others suffer big losses, rationales had to be developed for addressing equity between different groups of people—or not, as it turned out.

These three issues are tightly interrelated, though economists are quite deft at keeping them separate. Explaining the connection in theory quickly becomes complicated, but a simple example should help. Imagine two islands of identical size, with the same quality soils and rainfall patterns, the same number of tools and schools, and the same number of people. Everything about these islands is pretty much the same including the people's tastes for different goods. On one island, however, the resources—the land, tools, and access to schools—are quite evenly distributed, say rather as they are in Sweden. On the other island, 90 percent of the land, tools, and schooling go to 10 percent of the people while the other 90 percent of the people have just 10 percent of these resources, say as in Brazil. Now imagine that on both islands there are sufficient buyers and sellers for there to be competition, and that buyers and sellers are fully informed. Even though both economies are efficient in connecting what is produced with what people want, there will be significant differences between them. Incomes will range between lower and upper middle class in the island like Sweden, while in the island like Brazil there will be a few very rich, a small middle class, and many very poor people. With income distributions being so different, even though tastes are the same, the islands will produce different mixes of goods. There will be more beef and fine wine for the rich, and beans and rice for the poor in the Brazil-like island, while the Sweden-like island will produce chicken, a choice of starches, and vegetables for all. So a dam that produces irrigation for new grazing land for beef cattle will be valued much more highly on the Brazil-like island than on the Sweden-like island. The moral is that the values economists put into a cost–benefit analysis reflect the distribution of income, as much as the tastes of people. While it may be a minor difficulty when evaluating a small project, it is likely to be very important for a global problem.

For the most part economists set aside this distributional question as being too hard, and demanding judgments about the relative value of benefits and costs to the rich and poor that they were reluctant to make. Yet ignoring these issues meant that economic valuations and analyses tended to reproduce the status quo in the distribution of income, wealth, and power. Those who already were wealthy typically got more. So those who already had rights to land received irrigation water in the summer and flood control in the winter. Those who had no rights had no way to express what they wanted in any market behavior. Economists excused their inattention to the ethics of distribution across rich and poor by invoking platitudes about progress through economic development, wherein over time all would somehow become better off.

The association of cost–benefit analysis with development through environmental transformation continued through its application to poor countries by international development agencies such as the World Bank. Economic development guided by analysis that intrinsically favors the existing distribution of wealth and power is, however, likely to reproduce those inequalities, which in many countries have increased with development. Correspondingly, a method of analysis that reflects the existing distribution of wealth and power is equally unlikely to favor the types of changes needed to transition away from fossil fuels, deforestation, and other activities that drive climate change, precisely because it can favor the interests of those who prosper under existing arrangements. In this light, we see that the values that economists are using to estimate the costs of inaction are rooted in the economy we are trying to leave rather than the one we should be moving toward.

Economists have known that markets reflect the existing distribution of rights, income, and power since the early nineteenth century when the philosopher Augustin Cournot built a mathematical model of an economy (Cournot, 1838). This conclusion applies to the distribution of rights across generations too. If we care about future generations, current generations need to give future generations rights to a world with tolerable temperatures, less extreme storms, less rapid changes in sea level, and possessing an array of species and ecosystems that are not collapsing. Giving these rights to future generations means that cost benefit analysis should therefore embody a lower discount rate (Howarth and Norgaard, 1992).

When economists were first drawn into government through cost–benefit analysis in the Great Depression, they worried about what they could say given the importance of distribution. Since then, however, the profession has put a lot of effort into rationalizing the problem away, such that when a truly global distributional problem like climate change arises, with big implications for the distribution of rights across generations, economists lead us astray by using the wrong tools.

Economists were brought into policy analysis to help assure that political decisions were economically efficient by maximizing social benefits relative to costs (though as we have indicated, despite their best intentions, the way these benefits were valued was skewed). This has led to a professional culture of virtuous skepticism among cost–benefit analysts wherein they think of themselves as guardians of the public good against wasteful special interests who are happy to plunder the public purse for their own profit. As a consequence, they have sometimes tended to be as skeptical of climate scientists' warnings about catastrophic change as they have been of dam builders' promises about high social benefits and low project costs. We now examine the contemporary role of climate economists in a bit more detail.

WILLIAM NORDHAUS AND ECONOMIC OPTIMIZATION

Most economists think of decisions to mitigate and adapt to climate change as investments in a better future. In this light, investments to avert or adapt to climate change should compete with other investments in that future: in new infrastructure, new production facilities, and new commercial and residential structures. Surely, economists reason, there must be an optimal mix of investments in climate mitigation and adaptation and these other sorts of investments that best improves human well-being overall. Leading climate economist William Nordhaus has prominently defended this view. Following economic reasoning, if the rich who can afford to help really want to benefit others, there are plenty of opportunities. On this basis, Bjørn Lomborg (2007) argued that rich people should be "investing" in whatever has the highest returns for helping humanity, and that climate mitigation and adaptation have low returns compared to (say) controlling malaria or educating women in developing countries. If the world were to follow Lomborg's advice, future generations would be somewhat worse off due to climate change, but they will be much better off due to the other investments made in human well-being. Thus "investing" in climate mitigation should be avoided until more conventional ways of helping the poor are undertaken. There is a simple counter argument to Lomborg's line of reasoning: if we really care about human well-being, we should undertake *both* climate investments and other investments in human well-being.

Among economists, this kind of position about investments actually needs no defense. Instead, debate centers on the details of the estimates of benefits and costs of ameliorating climate change and how future benefits should be weighed against the costs of investing in mitigation and adaptation.

Nordhaus (1993) built a successful economic model along these lines incorporating key aspects of the interactions between the economy and the climate. He linked a simple model of climate dynamics developed by climate scientist Stephen Schneider with a simple model of economic growth that generates greenhouse emissions. Nordhaus labeled his Dynamic, Integrated, Climate-Economy model DICE, and made it publically available. Many researchers used the model, for it was easy to use to explore the implications of alternative assumptions. While it incorporated little of the complexity of climate dynamics or the ways economies actually behave, it captured the essence of the time that can pass between when emissions occur and when their consequences are felt, and the key choice between investing in mitigation and living with the consequences of climate change. The model could be used to calculate an optimal tax on carbon emissions (that could change over time) to maximize economic well-being through economic growth.

Nordhaus (1993) used estimates of climate damages that now look low because the range of effects of climate change was only beginning to be understood at that time. Conversely, early estimates of the costs of switching to renewable energy tended to be on the high side compared to how they now look. The combination of low benefits from mitigation, high costs of mitigation technologies, and choice of a relatively high discount rate led Nordhaus and other economists to argue that little should be done to avoid climate change. Optimal carbon taxes were estimated to start in the range of $10 to $30 a ton. Even as Nordhaus updated his analysis in subsequent years, his conclusions supported only modest action in the near term.

A central controversy concerns whether or not the discount rate for the benefits of mitigation should be different from other investments. For mainstream economists, discount rates, just like interest rates, should be determined in the market, and be roughly equal to the average returns on all investments. Climate economists eventually reached a consensus that the discount rate for thinking about climate investments should be pretty much the same as for conventional investments. In practice this yielded a discount rate of around 3 percent per annum.

A quandary generally not discussed by economists also deserves some attention. Past investments had relatively high private returns—around 3 percent per annum—in part because the economy was not sustainable; it was reaping current benefits by generating costs for future peoples through climate change. The past investment returns from which the discount rate was determined are therefore distorted by the fact that the costs of climate change were not considered. So, we have a problem of circular reasoning. If the costs of climate change are small, this circularity problem is not important, but what if the costs are large, as many climate scientists now argue?

Further, as philosopher Mark Sagoff has repeatedly noted (for example, Sagoff, 2011), investing in climate change mitigation is actually a bit different from private investment. While ordinary private investments are about helping ourselves over time, mitigating climate change is much more a matter of how best to help others in need. Investments in mitigation benefit future generations and, in particular, the poor and vulnerable in all generations.

NICHOLAS STERN, UNCERTAINTY, AND EQUITY

The "Nordhaus hegemony" among climate economists has not entirely disappeared. But more economists now join climate scientists to accept that uncertainty and the risks of extreme outcomes arising more quickly than previously expected should frame the debate. In this light, the problem is no longer one of determining an optimal investment path for climate mitigation, but rather more like building a home that will not collapse upon its occupants in a typical winter storm. One should build with a safe margin, especially when the costs of safety are not great and the costs of failure might be enormous.

There has also been an increasing awareness even among economists that climate change is ultimately a problem of equity (which we will address further in Chapter 5). Rich nations (and the minority of rich consumers in poor nations) have prospered economically through the burning of fossil fuels, while most of the damage will be felt by poorer nations. In an analysis that considers how economic activities between 1960 and 2000 of high-, middle-, and low-income nations impose ecological costs on themselves and on each other, Srinivasan et al. (2008) found that the effects of climate change had by far the biggest impact compared to other factors. (They also considered ozone layer depletion, fisheries decline, mangrove loss, agricultural expansion, and increased use of chemicals.) Their analysis of the "footprints" that the three kinds of countries impose on each other is summarized in Figure 3.1. The footprint of the rich countries on other countries is especially large.

Concern with the uncertainties of climate change and the inequities between rich and poor drove the landmark Stern Review of the economics of climate change commissioned by

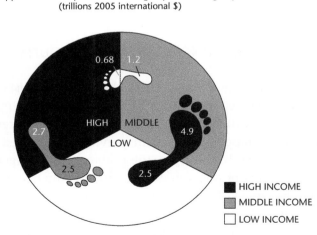

Figure 3.1. Where ecological footprints fall
Source: made by Uthara Srinivasan, reprinted by permission.

the UK's chancellor of the exchequer in 2005 as an independent assessment of the full range of economic issues raised by climate science (Stern, 2007). While Nicholas Stern had long been concerned about justice, no one can accuse him of being a radical critic of modern economics. He was head of the UK Government Economic Service at the time he was commissioned, and before that chief economist and senior vice president at the World Bank. Stern assembled a 23-member team of economists and climate researchers. Rather than ask what was the optimal response to a well-understood investment problem, the Stern Review asked what was the appropriate response, given uncertainties revealed by the science, to avoid, with a high probability of success, the imposition on future peoples of the more extreme dangers of climate change. Stern's team also worried about particular peoples, notably those living near sea level and those vulnerable to extreme climate events.

The Stern Review addressed risks and ethical issues that earlier economic analysts had hidden. When it came to discount rates in particular, Stern argued that ethics determined a low rate of around 0.1 percent per year. This low rate implies giving much more weight to long-term costs (and benefits) than the rate of around 3 percent per year used by conventional market economists. Unsurprisingly, Stern was berated by Nordhaus (2007a and b) for this choice. Stern determined that the value of reducing the emission of carbon to the atmosphere today was therefore US$390/ton, or some 10 to 20 times greater than the value determined by most economists before him. In Australia, economist Ross Garnaut (2008) produced a report for the national government reaching similar conclusions.

Given the big difference in their conclusions, established climate economists accused the Stern report of being weakly rooted in economic theory and empirical evidence compared to their own work (see, for example, Nordhaus, 2007a; a reply by Stern and Taylor, 2007; Mendelsohn, 2008). They argued that Stern used a discount rate that was too low, that he overstated the benefits of mitigation, and understated the costs; exactly the kind of results one would expect from an analysis that gives the poor and future generations more rights, which Stern did. They argued further that Stern paid too little attention to the development opportunities lost by investing in climate mitigation rather than in economic growth.

Stern was no radical. His review did not criticize the use of economics as a basis for reasoning about policy (and is, indeed, a very ambitious piece of planetary-scale cost–benefit analysis). His team accepted the importance of economic valuation, and did not argue for the intrinsic value of nature or that absolute priority should be given to the global poor, or to future generations. The Stern Review is in many respects pragmatic to a fault. To the consternation of climate scientists and biologists concerned about the ecological effects of even modest warming, Stern argued that, though 450 ppm CO_2 in the atmosphere was a reasonable long-term target, it was simply too difficult to stay below 450 ppm in the medium term given the material needs of the poor in developing nations. There are, as we have seen, scientists and even some economists who argue that a target of 350 ppm is morally and economically reasonable. In this light, the Stern Review is an improvement over the older climate economics but still deficient (Hansen, 2009).

CONFRONTING POTENTIAL CATASTROPHE

Conventional economics has a hard time coming to grips with potentially catastrophic outcomes that have a low but highly uncertain likelihood. Figure 3.2 captures scientists' concerns with these risks of not taking action on climate change. The figure portrays the probabilities (on the y-axis) of different climate sensitivities (on the x-axis) for a large number of different climate models. Climate sensitivity is an indicator of how many degrees the global climate will warm with a doubling of CO_2 in the atmosphere from the pre-industrial level (which was around 280 ppm). These models generally indicate that average global

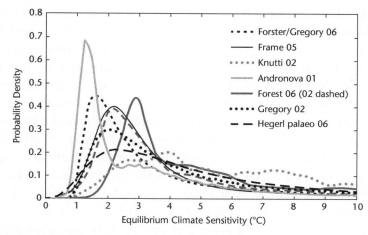

Figure 3.2. The fat tails in the risks of not taking action
The range of estimated probability density functions for the climate sensitivity from a variety of published studies. The y-axis represents probability, the x-axis the number of degrees centigrade warming that a doubling of CO_2 concentration in the atmosphere to 560 ppm would produce.
Source: derived from Intergovernmental Panel on Climate Change, Fourth Assessment Report, 2007, Volume 1, Box 10, Figure 2.

temperature will increase between 1.5 and 4.5°C, and this is the range that most economic analyses have considered in deciding whether it is worthwhile to invest in mitigation. Notice, however, that most of the models also indicate that there is a 1 percent to 5 percent chance—higher for a few models—that average global temperature will increase by a truly catastrophic 10°C or more. The fact that the curves do not converge to zero on the right-hand side of the diagram means the distributions have "fat tails."

The effects of a 4.5°C increase will be very significant while a 10°C warming would be absolutely disastrous for future generations. Looking at the probability distributions shown in this figure, economist Martin Weitzman argued that the uncertainty of our knowledge of the dynamics of climate change is such that the critical information actually resides in the "fat tails." Mainstream climate economists had actually been looking at the more likely outcomes and referring to them as extremes, rather than taking the real uncertainty and the really extreme possibilities seriously. According to Weitzman, the potential catastrophe accompanying the uncertainties of climate dynamics vastly outweighs any of the differences in valuation methods or discounting being debated by economists (Weitzman, 2009). Weitzman argues that benefits and costs would look a whole lot different if warming accelerated rapidly and a climate catastrophe unfolded (see also Sterner and Persson, 2008). The previously dominant Nordhaus-style climate economic analyses ignored outcomes that are as likely as the floods for which dams have long been built by the public sector, far more likely than the rare but catastrophic fire against which homeowners regularly buy insurance, and many times more likely than the terrorist attacks against which some countries now protect themselves to the tune of billions or even trillions of dollars per year.

ECONOMIC COSTS OR MORAL OBLIGATIONS?

The contention of early climate economists such as Nordhaus that mitigating climate change should be seen in terms of an investment choice has now been challenged many times over. Mitigation—and now adaptation—is centrally about helping or protecting others in different places and times. Howarth and Norgaard (1992) showed that when we choose to help or protect future generations, when we recognize their right to a climate more or less like our own, the economy can still satisfy social welfare. However it would follow a different path, with the prices of energy and material goods being higher (reflecting the real cost of fossil fuel use) and the rate of discount being lower. This makes valuation of costs and benefits on the basis of the prices we observe in today's profligate and unsustainable economy rather meaningless, unless you have already decided future generations and today's poor do not count. Climate economics proves not to provide definitive guidance about what to do. Economics and morality are intertwined, as any particular economic analysis proves to rest on some ethical (or unethical) assumptions, no matter how much economists try to hide them. In this light, the language of morality should not yield to the language of economics in political discussions (Jamieson, 2011).

Our judgments here should not be taken to imply that economic analysis should simply be dismissed. The trajectory from the early climate economists to their successors such as Stern and Garnaut shows the magnitude of the costs of inaction being progressively recognized and brought to public debate. As economics is stretched, it helps reveal what was previously just over the horizon. And, indeed, as evidence accumulates economists continue to adjust. Five years after his famous report that jump-started a new round of economic thinking, Nicholas Stern spoke out again, declaring that all the evidence now indicated that climate change was turning out to be a much bigger problem sooner than he had expected, that his report was too conservative (Stern, 2013). Economics is one starting point for coming to grips with the content, scope, and magnitude of the costs and risks of inaction, but economists should not be allowed to have the last word, for they are in the process of learning and rethinking themselves.

The intellectual struggles, deploying the tools of modern natural science and the blunter instruments of economics, are as we have seen intertwined with ethical issues, yielding what Stephen Gardiner (2011) calls "a perfect moral storm." Society and its experts remain "climate-challenged" when it comes to reaching collective judgment about the costs of inaction, understanding how these costs intermingle with moral obligations to humanity as a whole, and then taking action accordingly. But as we have also indicated, there are ways to contemplate what should be done in intelligent fashion, and we intend to build on these in the chapters that follow.

4

Actions that Promise and Practices that Fall Short

Almost all national governments now recognize the reality of climate change, and the need to respond. As should be clear from the previous chapter, the costs of inaction eventually become prohibitive. The repertoire of actions available to governments (and others) is substantial, and in this chapter we take a look at what can be done. Many actions ought to make a difference. Yet there prove to be formidable reasons why governments often do not adopt them; and when they do, policies that ought to work on paper are crafted and implemented in ways that render them less effective or even counterproductive. The reasons have a lot to do with the way powerful interests, dominant discourses, and political-economic systems are configured in today's world. They have still more to do with the profound and novel challenge that climate change presents—driving home the need to contemplate bigger questions about how societies are organized, not just what governments and others should do. These larger questions receive our attention in subsequent chapters, but it is important to examine the repertoire of available actions, still needed in any reconfigured systems.

Policy discussions often focus on major actions like a carbon tax or emissions trading scheme that would increase the cost of burning fossil fuel and so provide incentives to reduce its use, develop renewable technologies and, eventually, change lifestyles. But before rushing to design some optimal single instrument like this, we should think about all the other established practices and policies that make a difference, and that could be changed for the better. For example, coal mining is typically taxed lightly, but could be taxed more. In the US, homeowners can deduct mortgage interest from their taxable income, encouraging construction of large homes spaced further apart; that deduction could be focused on more efficient housing. Zoning laws could be changed to make urban landscapes more energy efficient and pedestrian friendly. Governments could redirect existing spending on research and development to cleaner energy. Further action possibilities include sequestration and long-term storage of carbon in forests and other biomass, and discouraging the release of carbon from plants resulting from land clearing. While economists in particular often speak as though we can and should design an optimal carbon tax or emissions trading scheme, in practice, reality will necessarily involve multifaceted action and continuous adjustment (just as in most policy areas).

Actions responding to climate change are normally divided into "mitigation" and "adaptation." Mitigation is directed at curbing the effect of greenhouse gases on the climate (especially through emissions reduction), adaptation at coping with the consequences of climate change. We begin with policies for mitigating by controlling emissions, before turning to other sorts of mitigation, and then to adaptation. Emissions control policies have been the subject of a lot of analysis and advocacy, and some implementation. But as we will see, a lot of the advice governments get assumes the world works in ways that do not reflect reality.

MITIGATION: REDUCING EMISSIONS

Total emissions of CO_2, the most significant greenhouse gas, depend on the size of the economy and its carbon intensity—that is, the amount of CO_2 emissions accompanying a unit of economic output. Contemporary political and economic systems seem to be addicted to economic growth, making it hard to operate on the size of the economy. Nevertheless, there are now serious scholarly and pragmatic arguments being advanced in an effort to understand how rich nations could transition to an economy that delivers security, community, and happiness while lowering material and energy use that we will address in the final chapter. The size of the economy could also be restricted by population control, pursued most seriously in China, but too controversial elsewhere (and increasingly in China as well). Thus, for now, the path of lesser resistance involves working on the greenhouse emissions intensity of the economy—though even if the economy were not growing, emissions would still need to be reduced.

CO_2 emissions can be curbed by increasing the efficiency with which fuel is used, or by switching to wind and solar energy, biofuels, and nuclear power. This transition is underway in many countries, but not proceeding fast enough to reduce emissions given rates of economic growth. Carbon intensity can also be reduced by producing services rather than material goods, and by lifestyles that involve living more compactly in cities rather than commuting from spacious suburban homes.

Other greenhouse emissions such as methane, aerosols, and particulates also depend on size of the economy, yet their levels do not have to vary proportionately with that size. Switching from coal to natural gas reduces carbon emissions but risks higher methane releases through leaks from gas fields and pipelines. Paddy rice production and cows also produce significant amounts of methane. Black carbon and other particulates are tied to the combustion of coal and diesel oil, some industrial processes, wood burning, and wildfires. In short, many phenomena associated with economic activities and land management affect the climate, some of which can be curbed more readily than others.

POLICY INSTRUMENTS FOR REDUCING EMISSIONS

Among the most popular policy instruments available in a market economy are emissions trading, otherwise known as cap-and-trade, and a carbon tax (Nordhaus, 2007b). Both provide incentives for producers and consumers to reduce emissions, and both should stimulate behavioral and technological change to conserve energy, or produce it from renewable sources.

An emissions trading scheme involves placing a cap on total emissions allowed for an economy or set of activities, then dividing that total into a number of permits. The initial allocation of permits can be done through auction or simply by allocating to existing emitters (especially companies). Thereafter, the permits can be bought and sold, and their price determined in the market. All this is overseen by some authority—normally a government, though some trading schemes (for example, the now defunct Chicago Climate Exchange) have arisen without government involvement. Emissions trading ought in theory to be the most efficient way to achieve any given level of reduction, because those for whom it is expensive to reduce emissions will buy permits from those for whom it is cheaper to cut emissions.

A look at policy practice tells a murkier story. The largest CO_2 emissions trading scheme to date was established by the EU in 2003 (Jordan et al., 2010). The EU scheme was widely hailed as exemplary but the price of permits crashed after the global financial crisis of 2008, which led to a drop in industrial production and so fall in demand for permits. This meant much of the potential of the scheme to reduce emissions was lost. In the long run, the positive effects on the global climate of that European slowdown in production are miniscule—suggesting that if the European scheme could be derailed by the emissions that were no longer needed as a result of the slowdown, it wasn't tough enough to begin with. A few countries have adopted emissions trading, and ten states in the Northeast of the US in 2008 established a scheme called the Regional Greenhouse Gas Initiative.

International CO_2 trading is also encouraged under the 1997 Kyoto Protocol.

A carbon tax puts a price per tonne on emissions of CO_2, so polluters then have an incentive to cut back, to reduce the tax they pay. Carbon taxes have been introduced in the Nordic countries, Netherlands, UK, and Australia; China plans one for 2015. In economic theory this is less attractive than emissions trading because it is not certain how much reduction any given level of tax will induce, so that level will need to be changed by trial and error. However, a look at the political practice that accompanies emissions trading schemes might lead to a different comparative judgment. While economists advocate that permits be auctioned, established polluters fight hard to receive free permits at the start of the scheme, reaping windfall profits as they can benefit from increased prices to consumers or by selling the permits on (Spash, 2010), and gaining a competitive advantage against more efficient, less polluting newcomers. A carbon tax may then look simpler, but it too is vulnerable to distortion by powerful interests seeking exemption or compensation (Stavins, 2009). For example, in Australia the entire agricultural sector was exempted from the carbon tax adopted in 2011; and the coal, steel, and coal-burning electricity industries successfully lobbied for compensation. Comparisons of emissions trading versus a carbon tax in terms of economic theory are in the end not decisive, because everything depends on the power politics accompanying their implementation.

This kind of power politics can be illustrated further through reference to a seemingly clever form of carbon trading to reduce emissions, organized by the Clean Development Mechanism (CDM) under the Kyoto Protocol. The idea here is that rather than cut back its own emissions, a wealthy country pays a poorer country to undertake a project in lower-emitting fashion than had been planned: for example, to construct a hydroelectric power station rather than an inefficient coal-fired one. Given the possibility of fraud, this requires verification—not just of what is actually done, but what was originally intended as a high-polluting alternative. The CDM can fund dubious practices. Notably, the price the CDM put on avoiding the emission of HFC-23, a byproduct from HCFC coolant production, was so high it meant that it became highly profitable for mostly Indian and Chinese factories to produce HCFCs (themselves contributing to ozone layer depletion) just so they could sell the credits for destroying the byproduct. As of 2012, HFC-23 produced by 19 factories accounted for 43 percent of the total trades under the CDM (UNEP, 2013). When the overseers of the CDM noticed the problem and moved to cut the price put on reducing HFC-23, some of the owners of the 19 factories threatened to vent the gas into the atmosphere (*New York Times*, 2012). This case drives home the problems with complex schemes whose fine print creates perverse incentives and vested interests that, once they are making money out of the scheme, battle to ensure harmful practices get perpetuated.

Though conventional economists are less fond of them, a host of regulatory measures can also be deployed. Regulation involves a government setting standards for polluters, who are fined if they violate those standards. Regulation can be "end of pipe" in simply specifying the quantity of allowable emissions by a polluter. It can also involve rules for allowable production technologies, the fuel efficiency of cars, building insulation, or types of heating and lighting. Regulation predates emissions trading and green taxes as the main instrument governments have used to control pollution, though any application to climate change was a long time coming. In the US, it required a lawsuit brought by the State of Massachusetts and others to force the federal Environmental Protection Agency in 2007 to recognize and regulate CO_2 as a pollutant. The climate policy laid out by Obama in 2013 is primarily regulatory.

Governments can also pursue mitigation by investing in infrastructure that is less polluting (bicycle paths and railways over roads and air), planning (to reduce urban sprawl), and subsidies and tax breaks for practices that either curb energy use or switch to renewable energy (such as housing insulation and rooftop solar panels). Governments can also reduce emissions that result from their own activities (such as running schools and hospitals, or heating office buildings). Governments could conceivably pass legislation to enable those suffering the negative effects of climate change to seek damages from those who caused it (though this would probably have to be a matter of international law). In many countries electricity is generated by publicly-owned industries, and so government can choose what kind of technology to use. Nuclear power remains an obvious (but controversial, in particular after the 2011 Fukushima disaster) way to reduce greenhouse gas emissions.

Individuals too can mitigate emissions by changing lifestyles and consumption decisions: to fly less, to commute a shorter distance to employment, to purchase products that are not carbon-intensive (such as locally grown fruit and vegetables). Governments and producers can help consumer decisions by requiring or choosing to label products; or by publicizing the emissions of polluters. Cultural change might make polluting practices less acceptable. Just as it is now socially unacceptable to smoke in many countries, it could conceivably become unacceptable to behave in climate-unfriendly fashion.

Corporations as well as consumers can choose to control emissions, even in the absence of governmental action. This might be a matter of cultivating a green image good for business, but it could also be that "pollution prevention pays" quite directly—if a company cuts emissions, it also cuts fuel and materials costs. There may be money to be made in developing and selling low-emission technologies. Companies can also choose to be active in emerging voluntary arrangements for emissions control. A more radical environmental role for business is implied in the 2012 remarks of World Business Council for Sustainable Development President Peter Bakker. Referring to environmentally recalcitrant businesses (not just in the context of climate change), he said "the 20 percent of really bad guys we need to regulate out of existence" (*Guardian*, 2012)—though it would have to be government doing the regulating.

TAKING CO_2 OUT OF THE ATMOSPHERE: BIO-SEQUESTRATION AND GEO-SEQUESTRATION

The mitigation story does not end with limiting emissions that go into the atmosphere; it can also entail taking greenhouse gases, in practice just CO_2, out of the atmosphere. The only proven practice to date involves growing vegetation, which at least while standing constitutes a carbon sink. The idea of *offsets* involves emitters paying for an equivalent amount of CO_2 to be absorbed by someone, somewhere planting trees. Generally it is developing countries that have sought to cash in—for example, under the Reducing Emissions from Deforestation and Forest Degradation (REDD) scheme operated under the auspices of the UNFCCC. There are however a number of problems with forest-related offsets. The first is that there is no guarantee the trees will actually get planted. The second is that the same plantings may be sold more than once. The third is that plantations of fast-growing trees may displace other land use activities, possible replacing complex ecosystems with simple ones, and ejecting traditional users of the land. The fourth is that it is not always clear that what is being paid for would not have happened anyway. The fifth is that trees do not last for ever: when they are cut down or burned, the carbon they had stored will sooner or later be released back into the atmosphere. The sixth is that the availability of offsets may be seen by polluters (including consumers) as an easy alternative to reducing emissions, and so may actually increase total emissions by removing guilt associated with them. Certification schemes for offsets can help solve the first four problems, but not the fifth and sixth. Forms of bio-sequestration that last longer than trees—such as conversion of organic matter into charcoal for incorporation into soils, "biochar"—look more promising in overcoming the fifth problem, but not the sixth.

A potentially longer-lived form of removal of CO_2 from the atmosphere involves carbon capture and storage, or geo-sequestration. The idea is to capture CO_2 emissions before they reach the atmosphere and inject them into underground reservoirs (such as exhausted gas fields), or possibly into the deep ocean (Meadowcroft and Langhelle, 2009). Obviously this only works for large stationary sources of CO_2, such as oil-, gas-, or coal-burning power stations—and only in particular places. Considerable uncertainty remains as to whether there is any secure and cost-effective way to lock up CO_2 in perpetuity underground or underwater; no large-scale example yet exists.

The capacity of the oceans to absorb CO_2 could be increased by increasing their iron content—a form of "geo-engineering." One very large problem is that this would increase the acidity of the oceans, with major ramifications for marine ecosystems. We will discuss other forms of geo-engineering (especially those intended to block solar radiation) in Chapter 7.

MITIGATION IN PERSPECTIVE

The measures we have discussed are all vulnerable to subversion, be it a heavy polluter using political power to get special treatment in an emissions trading scheme or compensation for a carbon tax, "greenwashing" of company products, corporations substituting public relations for action, consumers focusing on small decisions (recycling) while ignoring big ones (energy-intensive lifestyle). If greenhouse emissions associated with electricity generation have been reduced, it may be because of a switch from coal to gas. Gas produces lower emissions per kilowatt of electricity generated than does coal; but it is of course still a fossil fuel, so problematic in the longer run.

Now, the mere *possibility* of subversion is nothing to worry about; such is life in an imperfect world, in any policy area. More worrying is the degree to which subversion is endemic to contemporary political-economic systems as they confront climate change. The problem characteristics we identified in Chapter 1 enter in powerful fashion. The size of the material stakes and the lack of clarity in immediate and direct benefits of actions mean there are strong incentives for corporations and other interests to bend policies to suit their own profits—and equally powerful incentives for politicians, regulators, and publics to look the other way when they do. This is especially true when instruments are themselves complex (particularly the case with trading schemes and offsets). There are also powerful psychological and social incentives for citizens and consumers, who in principle accept the need to act, to displace that recognition from affecting their lives (see Chapter 2). All these mechanisms are reinforced by inertia facilitated by resistance (however unconscious) to the challenge to the idea of human progress that climate change represents.

In this light, it is no surprise that the instruments we have examined have to date only been adopted in piecemeal and limited fashion; their cumulative impact on total global emissions is negligible. However, the picture is not totally bleak. By adopting a combination of these measures some countries have made some progress in limiting emissions growth. The Netherlands is one of the most ambitious countries, the idea there involves coordinated, planned response including government, business, and experts, using multiple instruments. The Dutch government is committed in principle to "transition management" that envisages conversion of the national economy to sustainable, renewable energy by 2030 (see Kern and Howlett, 2009), though only in 2008 did Dutch CO_2 emissions begin to fall after a long increasing trend (Hajer, 2011: 13). The Dutch case offers at least a glimmer of hope: political systems are not equally problematic in their capacity to adopt and implement effective actions. We will pursue this question of how different kinds of governance either enable or impede action in Chapter 6.

ADAPTATION

Adaptation was left undefined by the UNFCCC in the original agreement on climate change in 1992. As Burton noted soon thereafter, adaptation was "an unacceptable, even politically incorrect idea" (1994: 11), in part because to focus on adaptation could be seen as avoiding the more difficult task of reducing greenhouse gas emissions. Mitigation continued to be prioritized in international negotiations, and in more local responses. But given the failure of mitigation, adaptation had to become more fashionable.

Adaptation is defined by the IPCC as "adjustment in natural or human systems in response to actual or expected climatic stimuli, or their effects, which moderates harm or exploits beneficial opportunities." A vast range of needs may require response. In the area of health alone, the needs may involve protection against thermal impacts on workers and the elderly, infectious diseases spreading outside their traditional range, declining air and water quality, food insecurity, the mental health of farmers, and aeroallergens (Hanna, 2011). In agriculture, adaptation may require new crops or breeds of stock, land management to deal with more or less rainfall, and the control of nutrients as soils change. Coastal cities need to worry about sea level rise, more severe and frequent weather events such as cyclones, and cities in general may need to cope with reduced water availability (Hunt and Watkiss, 2011). River deltas, low-lying islands, and semi-arid lands turning to desert may generate refugees who need to find new accommodation and livelihoods. In short, every negative consequence of climate change demands an adaptive response, be it to rising sea levels and more frequent hurricanes in Louisiana or to malnutrition in the highlands of Papua New Guinea (Barnett, 2011: 270). Dramatizing the challenge they face, in 2009 the government of the Maldives (most of whose land rises only a meter or so above sea level) held a cabinet meeting underwater, with ministers wearing scuba diving gear.

As recognized in the IPCC definition, ecosystems as well as human systems will need to adapt to the consequences of climate change. Human interventions can help. So for example more extensive protected areas and corridors between them will better allow species to shift their range as local climates change (Chester et al., 2012).

Adaptation can be anticipatory (before the fact) or reactive (in response to impacts). Anticipatory adaptation alleviates the need for eventual reactive adaptation. Think of how much less the costs of response to 2005's Hurricane Katrina in New Orleans would have been with higher levees or proper storm preparation strategies and evacuations plans. Multiply Katrina many times and we can begin to understand the relative merits of anticipatory and reactive policies.

Adaptation received much less attention than mitigation so long as it seemed possible the world might act decisively to prevent the worst aspects of climate change arriving. Many scientists and activists felt that to plan for adaptation was to admit defeat. Those in climate denial for their part believe that discussing adaptation is tantamount to accepting the validity of climate science, and so suppress discussion. However, adaptation eventually came to receive significant attention.

Some proposals for adaptation are minimal—to the point of being comical. Stay in the shade and use fans to keep cool, for example. More serious analyses categorize strategies into retreat, accommodation, and protection. To illustrate, coastal areas threatened by sea level rise could consider plans for retreating to higher ground, accommodating their population on a shrinking base of land, or protection through sea walls and barriers. The managerial worldview that underlies this particular taxonomy of actions has its uses when it comes to assessing vulnerabilities and risks, then evaluating and comparing different technologies and behaviors, and so we will devote substantial attention to it here. Yet we should not lose sight of larger questions pertaining for example to how questions of social justice play out in adaptation policy or how adaptation could be put in the service of larger social transformation in a climate-changed society (Pelling, 2011), involving ethical and cultural change (Adger et al., 2009). We return to these larger questions in later chapters.

ADAPTATION THROUGH REDUCING EXISTING VULNERABILITIES

One major theme in adaptation involves reducing the existing range of vulnerabilities in society in order to build strength to confront the stresses of climate change. For individuals, communities, and nonhuman species and ecosystems that are already stressed face adaptation at a disadvantage. Climate change can both exacerbate existing vulnerabilities and bring altogether new ones. Consider for example food security. Many societies have people who go hungry or lack adequate nutrition. Climate change can make things worse for them. A single storm, or a prolonged heat wave or drought, can wipe out a crop and so the income and nutrition it brings. In the summer of 2012, an extensive drought hit the Midwest of the US—and its important corn crop, which feeds both people and farm animals. The crop fell by 17 percent, and food prices rose, moving more low-income people into food insecurity. This occurred even with some anticipatory planning, as farmers had been planting varieties supposedly better able to withstand heat and drought. People can only anticipate so much. In Australia many farmers are already socioeconomically disadvantaged, have poor access to health services, and feel the cultural expectation of being self-reliant and stoic (Berry et al., 2011). Drought and economic uncertainty already lead to depression and self-harm; if extreme weather becomes more common, matters get worse.

Climate change will, then, make existing health vulnerabilities worse (Blashki et al., 2011). The fear is not new diseases descending on an unsuspecting public, but instead the amplification of existing disorders and health inequities. Hanna (2011: 219) argues that the threat to human health from a warming climate comes largely from the increase in frequency and intensity of extreme heat events that can push the human body beyond physiological comfort and undermine the capacity to work and function. It is "the elderly, the socially isolated or immobile, the very young, and people with chronic diseases, who are considered most vulnerable." In the European heat wave of 2003, deaths were more likely for the elderly poor—with no air conditioning, and no place to go. In the 1995 Chicago heat wave fear of crime kept many poor elderly from even leaving windows open in the evening (Klinenberg, 2003).

In this light, it makes sense to think first about improving the capacity to cope of the most vulnerable. Such improvement would involve measures not usually classified under climate change or even environmental policy. Examples would include the redistribution of income to the poor (both within and across national boundaries), promoting education, and efforts to increase human security more general. Social capital, which is the density of ties in social networks, turns out to matter a lot. Individuals and communities that lack social capital are often poorly placed to cope with challenges presented by life in general (Putnam, 2000). Climate change is no exception: in the Chicago heat wave of 1995, it was isolated, poor, elderly bachelors—those without social connections—who were most likely to die. In this light, the key to adaptation may be social capital and the ability to act collectively (Adger, 2003). Vulnerability turns out not to be simple exposure to hazards, but also a matter of how people are placed individually and collectively to respond (Spickett et al., 2011: 11).

TOOLS FOR ADAPTATION PLANNING: RISK ASSESSMENT AND COST–BENEFIT ANALYSIS

The most popular approach by governments to adaptation planning has been risk assessment. Risk assessment entails establishing a baseline desired state of affairs, identifying and evaluating threats to that state, estimating their likelihood, developing policy options to reduce or prepare for threats, and implementing (or preparing to implement) such policies. Risk assessment concerning health, food security, or storm damage has become the standard first step for public health, agricultural, emergency management, and environmental agencies taking on the challenge of adaptation.

Climate change has brought extensive risk analyses of heat, drought, river levels, storm paths and intensity, sea level rise, health, and much more. The forecasting of risks can be informed by data collection about the existing trajectory of chronic change (such as sea level rise) and the frequency and intensity of extreme events (such as droughts or typhoons). There is always a level of uncertainty, but that does not prevent the rational allocation of resources to counteract the most likely and/or most damaging risks.

In order to decide where to allocate these resources, cost–benefit analysis can be used. We discussed cost–benefit analysis at length in the previous chapter in the mitigation context of assessment of the overall costs and benefits of trying to limit greenhouse emissions. Logically, though, any analysis of the costs and benefits of mitigation should also incorporate estimates of the costs of adaptation—felt to the degree mitigation is not pursued or is not successful. Indeed, in his landmark review Stern (2007) presented the high costs of living with and adapting to climate change as an argument for mitigation. Other economists have deployed cost–benefit analysis to argue the costs of adaptation may actually be less than the projected costs of preventing or mitigating climate change. To put it crudely, it may just be that building higher sea walls is cheaper than converting the planet to wholly new post-carbon energy systems. So Bosello et al. (2012: 37) suggest that if one strategy is to be privileged on purely cost–benefit grounds, it should be planning for reactive adaptation. Other economists suggest we use cost–benefit analysis to determine the proper ratio of mitigation to adaptation spending (for example, Agrawala, 2011; Fisher-Vanden et al., 2011). As the likelihood of successful prevention of climate change recedes with the hopes for a global treaty, and as the frequency of costly events such as hurricanes increases, the ratio begins to shift, justifying the allocation of more resources to the adaptive side of the ratio. Analysis of the cost-effectiveness of adaptation research and development, potential anticipatory adaptive investments, and preparations for reactive policies is therefore both prudent and crucial to government planning.

This kind of analysis is, however, quite difficult, because the basic estimates of costs vary so widely. In studies of US adaptation to health impacts alone, cost estimates range from $2 billion (Nordhaus and Boyer, 2000) to $10 billion (Hanemann, 2008). Variations in cost estimates reflect long-standing controversies about what kinds of costs to include and how to assign monetary value to them; some costs (for example, to the integrity of ecosystems, or to aesthetic values) are less readily incorporated than others.

WILLFUL NEGLECT AND MALADAPTIVE BARRIERS

Adaptation policies face familiar problems that arise in any policy area: inadequate information, lack of resources or competence, and a culture of reactive (as opposed to proactive) management (Measham et al., 2010: 10). Unfortunately, adaptation policy also faces some specific problems when it is undermined or even precluded by ideologically driven austerity or anti-science policy makers. Policies to address and alleviate existing vulnerabilities, to build social capital and so adaptive capacity, or simply to do basic risk assessment and disaster planning are all attacked by the same anti-environmental, anti-science, and austerity-driven thinking that in some places impedes mitigation.

In a few places (confined mostly to the US), the study of climate impacts and risks has been undermined by law. Notable here is a 2012 law passed by the Senate (but not the House) in North Carolina that prohibited state agencies from reporting that sea levels are rising faster than in the past. As an article in *Nature* put it, nature itself seemed to mock the legislators, with research identifying the east coast of the US as experiencing the fastest sea-level rise in the world (Phillips, 2012). According to legend King Canute once set his throne on the beach and ordered the tide not to come in. King Canute was much wiser than the senators of North Carolina, for he knew his order was futile (he was dramatizing the limits to his own power to flattering courtiers). Imposed ignorance about the risk of sea level rise threatens other coastal states in the US, including Texas and Virginia.

EASING ADAPTATION

Especially in the face of maladaptive barriers to effective action, "no regrets" policies become attractive. These are actions that would be desirable and beneficial even in the absence of risks imposed by climate change. Cinner et al. (2012) argue that adaptive capacity can be improved by poverty reduction, literacy, good governance—all things good to attempt anyway. Some no regrets policies are mitigation strategies as well—for example, insulating buildings or blocking solar heat gain of windows both prepares residents to withstand higher temperatures and lowers carbon emissions (Bosello et al., 2012).

Especially if they have such no-regrets features (but even if they do not), adaptation policies should be more feasible for national governments to adopt than is the case for mitigation policies for one big reason. Governments often fail to act on mitigation in the hope of a "free ride" on the mitigation efforts of other governments; or, failing that, hesitate to move first because others might sit back and reap the benefit. This free rider problem is much less of an issue in adaptation, as most, if not all, adaptation spending benefits only the national or local population.

Adaptation can also be linked to the idea of sustainable development (Pelling, 2011), which since the mid-1980s has been the dominant discourse of environmental concern in global affairs. This discourse stresses the compatibility of environmental conservation with economic growth, as well as social justice between rich and poor and across generations. So the equity component of sustainable development and adaptation both address the situation of the most vulnerable, and their capacity to act for themselves. Quite how this might be done remains contentious. The idea of sustainable development can support anything from economic growth as usual to targeted development of technologies appropriate for the rural poor in stressed environments in developing countries. The latter might benefit from financial assistance from the wealthier countries responsible for most of the emissions that caused the problem that leads eventually to the need for adaptation.

Adaptation is not just a matter of policy choice and planning informed by techniques such as risk assessment and cost–benefit analysis. Important governance questions also come to the fore. In particular, public participation in policy development in and for vulnerable communities can generate support for and so the legitimacy of adaptation plans and policies—as well as bringing local knowledge to bear in the content of those plans and policies. This kind of participation is further beneficial inasmuch as it develops social capital, ideally involved in a virtuous spiral of adaptive awareness, community capacity, individual competence, and preparedness. We will return to more reconstructive forms of adaptation in Chapter 8.

CONCLUSION

Mitigation, and adaptation to a somewhat lesser degree, operate in a territory of policy instruments that ought to work in theory, but have a rough time in political practice; of powerful interests that bend seemingly attractive instruments to their own material advantage; of unproven technologies, some with major side effects; of options that are feasible and attractive because they promise little pain, but also in the end may not make enough of a difference; of complex schemes that create perverse incentives. These problems are not a matter of bad luck, or failure of political will, or poor analysis in policy design. Rather, they are endemic to contemporary political-economic systems as they confront the particular kind of challenge that climate change presents. Responding effectively to climate change cannot therefore be a matter of simply choosing better policies—still less of choosing better policies on the assumption that the world is basically a market whose incentives need to be tweaked. The instruments we have discussed may be useful—but only in the context of political-economic systems that can overcome the pervasive problems we have identified. Just what these systems could look like is addressed in the chapters that follow.

5

What's Just?

WHY JUSTICE MATTERS

Climate change is a collective global issue that requires a collective response—but that is not its whole story. Those most responsible for the existence of problems (industrialized societies with a long history of emissions) are not the same people or countries as those likely to be most affected by the consequences. Nor are they necessarily the same as those most capable of doing something; so for example China has a capacity that is much greater than its historical responsibility. What this means is that major questions of justice arise in the distribution of the burdens and benefits of action (and inaction). Climate change crystallizes questions that pervade many global issues, and so provides a crucial arena for thinking about how to pursue justice in general.

How then do we use justice to frame social responses to a climate-changing world? Any response here is complicated by the fact that justice itself is what political theorists call an "essentially contested concept" on which agreement is not possible, even in theory, despite the philosophical industry that addresses the concept. Further, and strikingly in the case of climate change, the range of injustices experienced can become "compound" when existing ones make new ones even worse (Shue, 1992).

For two decades climate justice has been addressed and contested by negotiators, social movements, and philosophers. Justice means different things to different people at different stages of climate change. We examine definitions of justice in ethical theory, in its articulation in global negotiations, and for movements that demand it. While there is some general agreement on injustices in the causes and effects of climate change, justice claims can differ strikingly, and sometimes clash. We also explore the prospects for justice in adaptation to climate change.

THE ORIGINAL DIMENSIONS OF CLIMATE JUSTICE

Key dimensions of climate justice were identified in the original agreement of UNFCCC in 1992. This specified that nations would work to "protect the climate system for the benefit of present and future generations of mankind, on the basis of equity and in accordance with their common but differentiated responsibilities and respective capacities." That single line in the agreement entailed several assumptions about justice:

- That the climate system benefits and supports human life (so the role of the natural world in the provision of justice was acknowledged).
- That there is a responsibility of present governments to benefit both present and future human beings.
- That such responsibility extends to people beyond national borders.
- That equity is to be a basic principle or ethic of international climate agreements.
- That the responsibility to address climate change would be based, in part, on causal responsibility—different for different countries, even in the face of the ideal of equity.

- That the actions required of nations would be based not only on these different responsibilities, but also on their different capacities to act.

The Kyoto Protocol that ensued five years later attempted to implement the idea of common but differentiated responsibility by famously requiring the 37 developed nations listed in its "Annex One" to reduce emissions (in a period extending from 2008 to 2012), while postponing action on the part of the developing world. It is worth reflecting further on the 1992 assumptions as they initiated debate on the relationship between climate change, government action, and social justice that continues to this day. In particular, the idea of "common but differentiated responsibilities" represents at least two different ways of understanding climate justice, and we turn to these now. The first is a matter of correcting historical injustice, the second a way of looking at equality in the present.

HISTORICAL RESPONSIBILITY

One ethical justification for differentiated responsibilities, and so for initial action on the part of only developed countries, was that wealthy nations caused the bulk of the emissions that cause climate change. Thus these are the countries that should now bear the burden of correcting what their industrial development has done to the world. This historical responsibility approach to justice is based on a straightforward polluter-pays principle.

This way of understanding the (in)justice of climate change is commonly articulated by Southern and developing nations along with social movements and non-governmental organizations (NGOs) that advance the interests of the developing world (Agarwal et al., 2002; Ikeme, 2003). At the 2011 Durban Conference of the Parties, for example, Lidy Nacpil of Jubilee-South-Asia/Pacific Movement on Debt and Development insisted that "[a]ll countries and peoples must contribute to the effort to reduce global GHG [greenhouse gas] emissions. But those responsible for the crisis must bear the greater share, proportional to their historical and continuing responsibility for the climate crisis" (Climate Justice Now!, 2011). In other words, they need to pay their climate debt. These national governments, NGOs, and activists are joined by some philosophers and economists who make the same argument (for example Shue, 1999).

The countries identified as most responsible in the historical approach also continue to have the highest emissions, in per capita terms—the average individual in the developed countries still emits much more than the average person in developing nations. This approach also recognizes that those responsible are *not* those most vulnerable to the impacts of climate change, who are mostly in the developing world, sub-Saharan Africa being particularly vulnerable. Impacts are distributed inequitably. Finally, the historical approach acknowledges that the same developed nations that have benefitted from two centuries of unregulated emissions are now those most *capable* of acting, due to the advantages that their economic, technological, and governance infrastructure (built on a history of fossil fuel consumption) gives them.

Critics of the historical approach respond that it is too backward-looking, that the allegedly guilty nations did not until recently have any idea about the consequences of their actions for the climate. This counterargument only goes so far; we may not be able to blame past generations back to the beginning of industrialization, but the evidence has been compelling for at least the past 20 years (say, back to the original UNFCCC agreement in 1992), and plausible for much longer than that (see Chapter 2). Critics also say this approach is not politically realistic; clearly, developed countries and those most responsible for greenhouse gas emissions do not want to accept the entire responsibility—for either the mess or the clean-up—and are unlikely to agree to any plan that does not include action by developing nations. This was at the core of the withdrawal from the Kyoto agreement by the US under President George W. Bush in 2001, Canada's abandonment in 2011, and Australia's reluctance to extend the protocol beyond 2012. The critics can argue in ethical terms that letting developing countries do nothing would give them unfair advantage when it comes to terms of trade in the contemporary global economy, facilitating (for example) China's growth at the expense of the US. Some observers see any focus on justice as a hindrance to international agreement on climate change (Posner and Weinbach, 2010).

One limit to putting the entire burden on those who historically caused emissions is the argument that, in some way, everyone is now responsible. We all emit, and developing countries are currently the fastest *growing* emitters of greenhouse gases. The developed nations point to China, now the single largest emitting nation—and growing quickly. China avoided commitment in the Kyoto Protocol because of its lack of past emissions. Further, countries like Brazil, India, and China now have large populations of rich consumer-emitters whose historical responsibility is no less than that of their counterparts in (say) Germany and Japan; should they be allowed to continue their profligate ways by hiding behind their poor compatriots? History is not what it used to be, and the historical approach is not as simple as it seems.

CURRENT EQUITY

Exclusive focus on historical responsibility ignores the fact that we are all in this together now. A more forward-looking approach to justice would be for everyone to have some sort of responsibility in the present. The challenge is how we respond to the issue in the here and now, where justice may demand equitable distribution of responsibility.

One equity response is pretty straightforward. Let us simply determine the total we can now emit into the atmosphere without causing too much climate disruption. Once scientists (impartially) determine that total amount, divide the total by the number of people on the planet. Then everyone—no matter who, or where, their history, or what their emissions are for—gets the exact same allowance. Everyone gets an equal slice of the greenhouse gas emission pie, or an equal "share of the capacity of the atmospheric sink" (Singer, 2004: 43). To make the process more manageable (rather than monitoring every human being), each country would be allowed to emit the product of its population times the allowable emissions per person.

The fact is that some countries will use more than their share of equal emissions, and others will use less. So Singer (2004) supplements his per capita equity approach with a trade system, whereby countries that need more than their allowance can buy permits from those who emit less. This system should result in both lower emissions overall and compensation to those poorer nations who use less than the per capita share (though its would depend crucially on carbon markets working well—see Chapter 4 on why they might not). It would mean those with a greater share of historically responsibility would pay more.

This per capita equity approach developed in reaction to some of the problems in a purely historical approach, offering an alternative based in a simple and pragmatic conception of justice. It is interesting, however, that the approach has not been picked up or articulated by movement groups who may see it as a dismissal of the responsibilities for past actions. In addition, this per-capita equity-based approach to climate justice misses the fact that different people, in different places, may need more or less than others to construct and live a decent life. And here is where rights and needs have reframed the climate justice debate.

DEVELOPMENT RIGHTS, HUMAN RIGHTS, AND CAPABILITIES

Demands for climate justice are often articulated around expectations of the impact of climate change, and of policies that respond to it. It is the poor, present and future, who will suffer most from floods, droughts, food insecurity, disease, forced migration, from any policies that limit the development of poorer nations, and even from attempts to foster sinks for emissions (for example, fast-growing trees that displace forest dwellers and users). Addressing the basic needs and rights of those most vulnerable is crucial.

In the face of wealthy nations' refusal to accept historical responsibility, and their insistence that all countries contribute to limiting emissions, the development rights approach has articulated a conception of climate justice based in *"the right of all people to reach a dignified level of sustainable human development free of the privations of poverty"* (EcoEquity, 2008, emphasis in original). All peoples and nations would have a right to develop to a certain level—just 125 percent of the global poverty level—before taking on any responsibility for emissions reduction.

This idea follows the UNFCCC principle of common but differentiated responsibility, draws on rights, and recognizes a differentiation of types of needs. It embodies arguments made initially by Henry Shue (1993) with regard to the difference between "luxury" and "subsistence" greenhouse gas emissions. In wealthy countries (and for wealthy consumers in poor countries) emissions result largely from the production of luxuries. Those who emit in order to supply basic needs and subsistence should not be required to pay for, or limit, these emissions, especially while others are emitting to drive luxury cars or produce unnecessary consumer goods. This is one area where philosophical arguments and social movement demands overlap.

A development rights approach allows for subsistence emissions for basic needs without penalty, while seeking to charge for and limit luxury emissions. Of course, it would be difficult to measure exactly what every person is getting for their emissions—subsistence or luxury. The idea is to add another layer of justification for assigning responsibility to the historically heavy-emitting developed nations, and to make sure the right to development is not violated by any global agreement that requires action or deprivation for those below a development threshold. While one can complain that this approach in practice stresses the nation-state rather than the individual, the subsistence versus luxury distinction can also be made at the individual level to bring attention to luxury emitters in poorer nations (millionaires in China or India, buying cars or air travel) while acknowledging the poor in rich nations (for example, those in poverty in the US) (Chakravarty et al., 2009).

But there are other human rights, in addition to "development," that may be threatened by climate change. Simon Caney (2010) insists that all people have a right not to suffer from climate impacts that undermine their basic interests. Climate change can undermine basic human rights, including life, health, subsistence, and against involuntary displacement. Caney's claims are not radical. Even concerning health, Caney remains fairly minimalist. He does not specify a *positive* right to healthcare for all; rather, a *negative* right that "other people do not act so as to create serious threats" to our health (2010: 167). This minimal negative right is threatened by climate change. From this perspective, justice would simply require paying attention to existing human rights and global agreements, for climate change will undermine the rights we already agree human beings should have. The obligation of states toward climate justice becomes an extension of their existing obligations to protect basic human rights, so there is no need for any new treaty.

Another way of understanding the threat to basic rights or needs is a capabilities-based approach that focuses on the range of conditions, or capabilities, necessary for people to have functioning lives (Holland, 2012). This range includes human rights, but in the context of what people need to construct lives for themselves—what is necessary to "do and be" as human beings. The capabilities in question might include political liberties, freedom of association, economic facilities, social opportunities, transparency guarantees, protective security, economic and social rights, health, bodily integrity, education, emotions and attachments, practical reason, social affiliation and respect, and control over one's environment (Sen, 1999; Nussbaum, 2006). This framework has been used by the UN to design the Human Development Index, as well as the Millennium Development Goals, emphasizing subsistence, health, education, gender equality, and the environment in which all of these can be sustained. In this light, climate justice requires a response to threats to basic needs and rights, which the international community has already agreed to tackle in the form of the Millennium Development Goals.

ENVIRONMENTAL RIGHTS

It may be that we can address climate justice with reference to existing rights and capabilities, but climate change also opens a striking new dimension to the justice debate. As illustrated by the 1992 UNFCCC agreement, it may be that climate change has made the international community finally come to terms with the reality that all human rights and needs *depend on an environment* that can sustain them.

Along these lines, Caney, after initially arguing that all climate justice requires is protection of existing human rights, claims that we may also have a right to an environment that does not undermine basic interests and needs. So people have a right not to suffer from drought and crop failure, heatstroke, infectious diseases, flooding and destruction of homes and infrastructure, enforced relocation, and "rapid, unpredictable, and dramatic changes to their natural, social, and economic world." People "have the human right not to suffer from the disadvantages generated by global climate change"—a right to the environment that provides and sustains other rights (Caney, 2005: 768). Combining the approaches of Caney and EcoEquity, Steve Vanderheiden's (2008) notion of climate justice is based in both environmental and development rights. The development right is rearticulated as a right to have the basic environment in which human flourishing is possible; this includes a stable climate.

From a capabilities perspective, the point is to incorporate environmental conditions and systems into a conception of justice. Variations in climate and, so, environment, will affect what individuals are able to do with the resources they have. If reduced rainfall makes it more difficult to grow food, or if flooding uproots people from their homes, then climate change limits their capability to convert resources into fully functioning lives. If climate change produces a barrier to that functioning, then climate change produces injustice.

Holland (2008, 2012) argues that all capabilities depend on a sustainable environment. If the capabilities approach is about functioning, and we all depend on particular environments to support that functioning, then this approach must attend to ecological systems. Holland recommends a "meta-capability" of "Sustainable Ecological Capacity," the level at which ecological systems could sustain not only themselves, but also the other basic capabilities for human beings in the system at an "environmental justice threshold." So only if ecological systems continue to sustain the conditions that enable capabilities, will "the *ecological* conditions of justice" be met (Holland 2008: 328).

It is important to understand relationships between basic rights, basic human needs, and their dependence on environmental conditions. The natural world in which we are immersed, and which we affect, can no longer be invisible and ignored, but must be thoroughly incorporated into both the philosophy and provision of justice. This challenge does not come only from philosophy; environmental and climate justice movements also stress the human/nature relationship, realizing that basic needs such as health, shelter, and food security will be affected by environmental conditions and undermined by climate change.

FROM CLIMATE RIGHTS TO CLIMATE DUTIES

Defining exactly what is just in the context of climate change is simply the first challenge. What to do to bring about more just responses comes next, bringing us back to the question of who is ultimately responsible. The historical approach clearly links a conception of justice with a demand for action on the part of those most responsible. Rights-based notions do the same, though in different ways.

For all rights, there are corresponding duties. Yes, we have basic rights, but all individuals *also* have the burden, duty, and responsibility associated with protecting those rights of *others*. So if current emissions levels in your everyday life are unjust, you have a responsibility to stop that infringement of another's rights. As Bell (2011: 115) puts it, we have a duty "not to accept benefits that result from actions that violate someone's human right." So we have a "duty not to emit greenhouse gases in excess" of our quota. Those that do exceed their quota may "have a duty to compensate others for their loss. If we are violating another's rights, we have a duty to stop, or to pay compensation."

Some people and communities in the high-emitting nations feel such a duty, and have adjusted their lifestyle or practices to limit their own emissions. But does duty end with buying a more fuel-efficient car, buying local, limiting air miles, or purchasing offsets? Bell (2011) insists our duty to climate justice extends beyond our individual emissions and responsibility not to benefit from impacts on others. He believes there is also a general duty to promote fair and effective institutions for the protection of human rights, including development. The fact that this kind of argument would apply to just about any imaginable right held by anyone, not just those connected to climate change, means however that it seems to impose an unrealistic burden on everyone to be a super-citizen. But the point is that conceptions of climate justice extend the range of responsibility, from the individual up to states and global institutions.

THE EXPANDING REALMS OF JUSTICE—OTHER COUNTRIES AND FUTURE GENERATIONS

From the original UNFCCC agreement in 1992 through to the establishment of the Green Climate Fund in Durban in 2011, an international conception of justice has been assumed in the language of the negotiations, if not implemented in practice. Climate change is of course a truly global issue. Some philosophers (notably John Rawls) dismiss the idea that the responsibilities of justice extend beyond national boundaries, within which justice has traditionally been conceived (and maybe even applied). But climate change thoroughly undermines any such geographically and conceptually limited view of justice.

The necessary extension of justice to the international or global level has been conceived of in two very different ways (Baer, 2011). We can think individually about our own obligations to people beyond our borders (Pogge, 2002). What do I, living in a wealthy country that primarily uses emissions-intensive coal for energy, owe to another individual person in a poor island nation that may be inundated by rising oceans? Can I conceive of my own obligations of justice reaching beyond the borders of my own community or country, to reach distant others?

True cosmopolitans focus only on the rights and obligations of individuals; other thinkers retain the idea that states matter. As Baer (2011: 326) explains:

> [p]recisely because GHG emissions cause harm that crosses borders, and because of the asymmetry between the highest emitters and the most vulnerable, there is a near consensus among the philosophers who have written on the topic that considerations of justice do in fact justify the obligation of rich and high-emitting countries

to take responsibility for reducing emissions and paying for adaptation. Baer makes a conceptual leap here, from a cosmopolitan justification for individual responsibility to an international framework for states. The framework for justice becomes not cosmopolitan, but rather international—the relationship between states, or between citizens of different states as mediated by their governments. The duties are of one state to another, or of states toward international agreements and obligations. In particular, the question is the relationship between states most responsible for climate change, and states most vulnerable to the impacts of that change. This kind of international justice argument is troubled by the problem we noted earlier that there are large numbers of rich consumers in relatively poor countries who would avoid any obligation, and large numbers of poor people in rich countries who would be burdened with obligations. Though it might be easier to implement, as a matter of morality international justice is inferior to cosmopolitan justice.

Climate negotiations have always had an eye on the long term, which is where many of the consequences of past and present actions and decisions will be felt. It is our future selves and descendants who will either benefit from what we do now, or be hurt by an impaired climate and the reduced capacity of ecological systems to support human life. Intergenerational justice therefore looms large in a climate-challenged society.

Yet we find that in both theory and practice, concern for the long term is lacking. In theory, there are some philosophers who argue that we have no obligations at all to future generations (largely on the grounds that we do not know who exactly they might be and what they might want). Other philosophers argue the case for caring about future persons (for a review see Thompson, 2009). As we noted in Chapter 3, in economics, the future is, quite literally, discounted. Mainstream economists believe that just as markets work on the basis of interest rates that mean that a quantity of money today can grow with time, the reverse logic also applies: future benefits and costs should be discounted by an annual percentage rate. The landmark Stern report to the UK government in 2006 made a moral judgment that this discount rate should be low: of the order of 0.1 percent per annum (Stern, 2007). Speaking for mainstream economists, Nordhaus (2007a and b) replied that such moral judgments had no place in economic policy decisions, and that a market-determined discount rate of around 3 percent per annum should be applied instead. What this means is that for Nordhaus, future damages from climate change appear much lower when converted to their present value, thoroughly undermining Stern's case for comprehensive and early action to reduce greenhouse gas emissions.

There is a high discount rate implicit in politics as well. Many of the impacts of past and present decisions will be felt by future people to whom governments (including democratic governments who look little further than the next election) are not presently accountable. The current state of political and economic systems suggests that many governments care little about the present poor. Why should they prioritize the *future* poor, let alone the future poor in other countries, still less those in the future who will not necessarily be poor? While the 1992 UNFCCC agreement acknowledged future generations, the international community has yet to face up to that particular ethical challenge.

RIGHTS, CAPABILITIES, AND MOVEMENT DEMANDS

The philosopher John Rawls famously said justice is the "first virtue" of social institutions, and though he said little about them, it is also often the first claim of social movements—including those active on climate change. Climate justice social movements deploy the whole range of ideas about history, equity, rights, capabilities, and functioning; yet the main concern is how climate change will affect everyday life, and reduce peoples' ability to sustain themselves. Activists insist that the key to climate justice is protecting vulnerable communities, and promoting what it takes for them to function.

Movement groups stress several specific individual and community level capabilities. To begin with health, as the International Climate Justice Network (2002) puts it, the impacts of climate change will "threaten the health of communities around the world." A focus on public health was also the first policy recommendation of a key "Advancing Climate Justice" conference in 2009 (WE ACT for Environmental Justice, 2009). As we saw in Chapter 4, health effects can stem from factors such as increased heat, changing rainfall patterns, changes in the geographical distribution of pathogens, increased incidence of severe weather events, and change in the composition of ecosystems. The consequences can include disease, starvation, and impaired mental health (indications of which can be seen in climate-stressed agricultural communities even in a rich country such as Australia).

Climate justice activists also highlight the current lack of recognition of vulnerable cultures and communities, and of the threats to them. It is not just that some are more vulnerable than others; it is also that this vulnerability is not respected. Indigenous peoples and island nations share a concern about how climate change endangers traditional practices and cultures tied to threatened places, from the Arctic to the tropics. The island nation of Tuvalu's proposal at COP-15 in Copenhagen in December 2009 to make industrialized countries accountable for the damage they were doing to low-lying island states was never going to be adopted by the gathered nations. The point was to bring attention and recognition to the peril in the most affected places. Recognition also arises as a demand when it comes to cultural loss likely to come from climate diasporas. A precursor of things to come can be found with the dispersal of African-American communities following the devastation of New Orleans by Hurricane Katrina in 2005. Such issues highlight the importance of social recognition and affiliation as central capabilities, necessary for human functioning and so for justice.

Climate justice activists also call for accountable, authentic, and effective public and community participation. So the Environmental Justice and Climate Change Initiative insists that at "all levels and in all realms, people must have a say in the decisions that affect their lives. Decision makers must include communities in the policy process" (EJCCI, 2002). The

Bali Principles of Climate Justice (International Climate Justice Network, 2002) demand that climate justice:

> affirms the right of indigenous peoples and local communities to participate effectively at every level of decision-making, including needs assessment, planning, implementation, enforcement and evaluation, the strict enforcement of principles of prior informed consent, and the right to say 'No'.

Democratic participation is seen as a right necessary to protect communities.

SHIFTING GEARS: JUST ADAPTATION

The global failure to act on the mitigation of emissions means that climate change is coming whatever the world now does. It is therefore necessary to think about adaptation, be it to floods in Pakistan, heat and fires in Russia, drought and failed harvests in sub-Saharan Africa, and the increasing frequency and intensity of severe weather events. These impacts are not, and will not, be equally distributed across the globe. Some people are more vulnerable than others, some have a greater capacity to adapt (enabled by wealth, good governance, and social capital) than others.

Many of the approaches to justice laid out above were designed with prevention or mitigation in mind; most have been proposed as normative frameworks for global agreements within the UNFCCC process. They have less to say about the challenge of just adaptation that now confronts us. However, rights-based and capabilities approaches to justice can be applied to specific and local adaptation endeavors, as well as to global agreements embodying basic needs and rights (Schlosberg, 2012). One of the great challenges of adaptation is that different people, in different communities, in different environments, will be vulnerable in very different ways, so the very specific things threatened in diverse locations need examining.

In a just adaptation, communities would be engaged in discussions about local vulnerabilities, as understood by a variety of stakeholders. Rather than a top-down, expert-driven affair, democratic participation would become central to specifying what rights and capabilities actually mean in particular contexts. Communities need to be thoroughly involved in both mapping their own vulnerabilities and designing policies to shield them from, or enable them to cope with, climate shifts threatening their ability to function.

That said, localized discourses and perceptions of vulnerability might differ across social groups. Indigenous peoples, farmers, landless laborers, and tourism managers might have a different sense of exactly what capabilities are made vulnerable through climate change. So localized vulnerability maps of basic human capabilities and the changes in climate and environmental conditions that threaten them will be necessary. Participation and deliberation—capabilities themselves—can help us to understand and determine distinct and local environmental needs, simultaneously satisfying recognition capabilities. All this points to engaging people and their varied experiences and understandings of vulnerability and risks in democratic deliberation about just adaptation policy.

CONCLUSIONS

Gardiner (2011: 312) rightly points out that "the world's attempts to address climate change seem to have fallen far short of taking justice seriously." Initiatives such as the Green Climate Fund (designed to transfer money for mitigation and adaptation projects) confirmed at the Durban COP in 2011 show that justice concerns sometimes still make it through (though the Fund is organized in top-down fashion, and much remains to be seen about how it is implemented). While noble sentiments about justice were articulated in the 1992 agreement, especially ideas about common but differentiated responsibilities, with time justice arguments have increasingly been deployed by negotiators in ways that happen to serve their national self-interest. So for the US, justice means forgetting the past, instead focusing on future fairness, especially equality in the terms of trade; anything else means giving competitors such as China an advantage by imposing the economic burden of emission reduction on itself while China does not also commit. For China and the G77 developing countries, justice is historical, which puts the entire burden of emission control on wealthy countries. While all of this helps to bring the language of justice into disrepute, the ethical arguments we have surveyed remain, and justice is still at the forefront of social movement demands.

Justice does then remain a key challenge to climate-changed societies—how it is defined, how it is implemented, *if* it is implemented. As we move from global attempts to prevent climate change to more localized processes to adapt to it, justice remains such a challenge. Maybe the need to talk about adaptation at the local level, in terms of the impacts of climate change on everyday life, will lead to more concern with justice in adaptation policy, even a return of the idea of common but differentiated responsibilities. Despite problems at the level of multilateral negotiations, theorists and activists alike have increasingly articulated a pressing range of climate justice concerns. Climate justice could go a long way to making a climate- *and* justice-challenged world in general a more just place.

6

Governance

WHY GOVERNANCE MATTERS

It is all very well to contemplate what policies are likely to prove effective and just, as we did in Chapters 4 and 5, but that assumes we have some effective authority to put them into practice. So comparisons of (for example) emissions trading schemes against carbon taxes as ways of achieving greenhouse gas emissions reduction need some kind of body to craft the policy and implement it effectively. Sometimes that body will be a sovereign government, but sometimes the government is missing altogether or has only partial jurisdiction—for example, if a carbon trading scheme can extend across national boundaries. When government is missing we can however still speak of processes of governance—notably at the global level. "Governance" is a broader concept that allows for more fluid, informal, and transnational arrangements, though it can also include government as conventionally defined.

Global inaction on greenhouse gas emission reduction and inadequate national policies are sometimes blamed simply on an absence of political will. But a big part of the story is that if we can't get the structure and process of governance right, we are not going to get the policies right. So it remains remarkable that some high-profile proposals either ignore the governance question altogether, or treat it in simplistic terms.

For an example of ignoring governance altogether (doubly remarkable in that the person doing it is a political scientist) consider the policy proposals ranked by a group of experts assembled by Bjørn Lomborg under the auspices of his Copenhagen Consensus Center (Lomborg, 2010). Using cost–benefit analysis plus expert judgment, the top three solutions were:

1. Marine cloud whitening research
2. Energy research and development
3. Stratospheric aerosol insertion research

The first of these would require creating mists from sea water to create clouds that better reflect sunlight, the third injecting tiny particles into the atmosphere to simulate the cooling effects of volcanic eruptions. Though cast in terms of "research" the reason for this ranking is the anticipated net benefit of the anticipated policy (policies to reduce greenhouse gas emissions came bottom).

The first and third options currently have no authority capable of implementing or regulating the kind of action they require, which to be effective must last forever. It might be possible for nations to take unilateral action within or above their own territory (including offshore waters)—but it is also unlikely, given that many of the benefits from marine cloud whitening and aerosol insertion would accrue beyond that territory, and to the globe as a whole (in terms of average temperature reduction), as would any undesirable side-effects. The requisite authority is also missing when it comes to many proposals to advance climate justice—for example, the internationally tradable per capita emission rights proposed by Singer that we discussed in Chapter 5. In short, new kinds of policies seem to require new kinds of governance.

For an example of treating governance in simplistic terms, consider the declaration by James Lovelock, best known for his "Gaia hypothesis" that treats the Earth as a whole as a self-regulating system, that:

> Even the best democracies agree that when a major war approaches, democracy must be put on hold for the time being. I have a feeling

that climate change may be an issue as severe as a war. It may be necessary to put democracy on hold for a while. (*Guardian*, 2010)

This statement is highly controversial: it is not clear that authoritarian governments currently do or could do any better than democracies on an issue like climate change. It is also vague: is Lovelock suggesting every country in the world put democracy on hold? Who will do the holding? What about the need for global action? The problem there cannot be democracy, because there is no global democracy.

Clearly we need to move beyond such simplistic treatments. How then to think about governance? We should not assume that existing national governments are unproblematic structures and just need supplementing when it comes to the international or global aspects of policy. As we pointed out in Chapter 1, existing social, economic, and political systems were not designed, and did not evolve, to cope with an issue like climate change, and that includes national governments. We should be alive to the possibility that climate change requires radical reworking of the very nature of governance. In this chapter we look at the range of governance questions, from the local to the global, and from familiar established institutions to reformed kinds of governance, though we put our analysis of truly innovative governance in the discussion of Chapter 8 on transition. Let us begin with the most familiar kind of governance, the national government of sovereign states.

NATIONAL GOVERNMENTS

The world has 195 states, ranging from mega-states such as China, India, and the USA to micro-states with only a few thousand inhabitants. Can we say anything about which particular states, and which kinds of states, do better or worse when it comes to dealing with the challenges presented by climate change?

The most straightforward performance indicators here are levels and rates of change of per capita greenhouse gas emissions (though matters are complicated a bit by international trade: emissions are usually charged to the country that produces rather than consumes goods, meaning importers can offload emissions to major exporters of manufactured goods like China). Generally, as income per capita rises, so do emissions—though it is not a simple one to one relationship, as countries at the same level of income per head can vary substantially in terms of emissions per head. As of 2005, the three highest per capita emitters were the US, Australia, and Canada, all of which also did particularly badly in terms of percentage increase in emissions between 1990 and 2007 (Christoff and Eckersley, 2011: 435). Germany and the UK did particularly well in reducing emissions between those two years, the former by 17.6 percent per head, the latter by 9.5 percent per head. Post-communist countries did even better but that was mostly a result of their economic and industrial collapse in the 1990s (Germany's apparent success also benefited from this effect as East Germany's emissions dropped massively after reunification with the West). China, India, and Brazil increased emissions heavily from a low base. France and Japan achieved substantial reductions before 1990 as a result of adopting nuclear power.

While there are variations in national performance, it would be hard to say that any country has done enough. Most of the Annex One wealthy countries that ratified the 1997 Kyoto Protocol failed to meet their 2012 targets. Moreover, even if they had been met those targets were only going to make a very small difference to the global trajectory of atmospheric greenhouse gas concentrations.

There are two reasons why states are for the most part currently not delivering when it comes to emissions reductions. The first is simply a collective action problem: each national government has an incentive to let others take the lead and bear the burden, while reaping a share of any benefits in the form of a more stable climate. We see this in international negotiations, where countries such as India and (until recently) China argue that all of the burden should fall on wealthier countries. We see it in national politics, where everywhere except the mega-states, opponents of action can argue that what their own country does will make little difference to the global total, so they might as well not bother, or at least wait until larger countries act.

The second reason for non-delivery is that developed states are the product of several hundred years of history in which they just had to perform particular functions, otherwise they would perish. The original functions were keeping domestic order, maintaining security in a potentially hostile international system, and raising revenues to finance these tasks. With time, their core functions expanded to encompass the promotion of economic growth (which helped when it came to raising revenues and security), and eventually systems of welfare and social security (which helped reduce the potential for domestic uprisings). Environmental conservation has never been a core priority of any state in this sense because it seemed states could survive quite well without it (Dryzek et al., 2003).

Yet some states have done better than others in attaching environmental concerns in general and climate concerns in particular to their traditional core priorities. One link that could apply in future would see the environment as a matter of national security; Gilman et al. (2011) highlight the security threats that a world disrupted by climate change could pose to the US. In addition, reducing greenhouse gas emissions could be seen as promoting energy security, as the US would not need to rely on oil imported from the unstable and problematic Middle East. Small island states and low-lying countries such as Bangladesh do not need to be told of the threat to their existence that climate change poses, but their political energy is likely to go into urging global action by others and adapting to the consequences of climate change, rather than domestic initiatives to curb emissions.

A link has been made to the core priority of economic growth via the idea of ecological modernization, associated with the slogan "pollution prevention pays," mainly because low pollution means efficient materials usage and a healthy workforce. To date this discourse of ecological modernization is most prevalent in Northern Europe and Japan. It tends to be associated with consensual political systems and "cooperative market economies" (Hall and Soskice, 2001) featuring negotiated understandings between business, labor, and government—which can be extended to other social interests such as environmentalism. This cluster can be contrasted with the politically adversarial and more competitive "liberal market economies" as epitomized by the Anglo-American countries (US, Canada, Australia, New Zealand, UK), where economy and environment appear in conflict, as opposed to the harmony of ecological modernization. It is the consensual/cooperative market economies/ecological modernization systems that have done best in limiting greenhouse gas emissions; and the adversarial/liberal market economies that have done worst—with the important exception of the UK. Things have changed since the 1970s, when the US led the world in environmental policy innovation.

In the UK, long an environmental laggard, both Labour and Conservative-Liberal coalition governments have since 2000 adopted ambitious targets for reducing greenhouse gas emissions. In 2011, Conservative Prime Minister David Cameron affirmed a commitment to reduce UK emissions to 50 percent below 1990 levels by 2025, and to an 80 percent reduction by 2050. Particularly striking in the UK is the fact that the large right-wing party argued that it had accepted the reality of climate science and actively sought environmentalist votes; in the 2010 election, their slogan was "vote blue, go green" (blue being the Conservative Party color). Clearly, their policy implementation fell short, and there is a question about the authenticity of this position. In stark contrast, however, emphatic denial flourishes unabated in the right-wing parties of the US, Canada, and Australia.

What happens in the three mega-states is going to matter enormously on the global scale. The US, with the largest historical emissions (i.e. accumulated total) of greenhouse gases, seems to face a perfect storm of impediments to effective action. To begin, it is an adversarial system with a liberal (competitive) market economy where the idea of ecological modernization has made limited headway. In Chapter 2 we pointed out that it has adversarial, legalistic mechanisms for achieving closure on policy issues—but the sheer number of veto points means losers can come back and try to find somewhere else to fight (a different court, another committee of Congress, a different level of government). Organized and emphatic climate change denial is more prevalent in US society than anywhere else, and is even fed by the balance doctrine in the serious media. Fossil fuel-producing states are over-represented in the US Senate, making ratification of any global agreement on climate change highly unlikely. US recalcitrance suggests the rest of the world may have to move ahead on climate change without the US federal government, for any foreseeable future. Despite the potential threat to national security that climate change poses, and despite the economic benefits that could accompany both energy efficiency and renewable technology development, entrenched power structures and interests have kept the US from adopting any national climate policy—even when, it seems, such action could serve the core priorities we discussed earlier. Some states (notably California) and local governments have stepped into the gap, but their efforts hardly compensate for the lack of national action. And other states feature disastrous climate politics, because extreme right-wing Republicans have demonized Agenda 21 (adopted at the 1992 United Nations Conference on Environment and Development) as an attempt to impose world government on the US. Because Agenda 21 was adopted in the name of sustainable development, that leads legislators in states such as Arizona to resist even minor steps to improve energy efficiency—for example, restrictions on the sale of incandescent light bulbs. The same imagined threat explains opposition to renewable energy and decentralized energy generation.

China, now the world's largest annual emitter of greenhouse gases, is as suspicious as the US of any international treaty, and still more protective of its national sovereignty. Yet China is also attractive to climate change authoritarians (for example Beeson,

99

2010) because its government seems capable of acting quickly and decisively, having accepted the reality of climate change. This capacity is especially apparent when it comes to introducing new technologies; from a very low base, China rapidly became the world leader in renewable energy technology. But China is also heavily dependent on inefficient coal-burning power stations, which it continues to build, and so its greenhouse gas emissions are increasing rapidly. China perhaps illustrates the inability of authoritarian systems to cope with multifaceted, complex problems associated with greenhouse gas emissions from multiple sources—introducing new technologies is by far the simplest part of the puzzle, and that is all China has been able to do so far (though in 2012 it began experimenting with a localized emissions trading scheme). Given China is playing "catch up," it is actually more straightforward and less expensive for it to pursue and adopt renewable technologies than it is for more established developed economies, which would have to take older sources of power offline. To date China has seemed uninterested in seizing the opportunity for constructive global leadership on climate change issues, unwilling to step into the vacuum left by the paralysis of the US.

India in late 2011 gained global attention by leading the opposition to any semblance of constructive global action at the Durban COP, insisting that richer countries had to take the lead. India remains committed to the historical responsibility argument as the essence of global justice (see Chapter 5). Domestically, India shows fewer signs than China of taking climate change seriously. Its economic growth is still fed by coal, and that growth is clearly the central imperative of the state. That said, India is expanding its development of sustainable energy alternatives (Bhattacharay and Chimnoy, 2009), and even has a Ministry of New and Renewable Energy.

When looking at comparative national performance we should not focus solely on government and the kind of economy a country possesses. The role played by social movements, activists, NGOs, and the media is also important; together these help constitute a larger "public sphere" that is a source of both ideas about what to do and pressure on governments actually to do it. One of the reasons Germany does comparatively well, on climate change no less than other environmental issues, is the historically energetic green public sphere that first flourished in the 1970s, and continues since then. This public sphere generated radical environmentalist critiques but also ideas about ecological modernization, becoming a factor that had to enter the electoral calculations of the established political parties, all of which were happy to take on green ideas when it suited them (Dryzek et al., 2003). More recently, we see this same trend in the UK, with even the Conservative Party claiming to be green. Comparing his country to the US, Frantz Untersteller, Environment Minister in the state of Baden-Württemberg, said in 2011 that one big reason Germany was able to move to decarbonization of its economy was that it didn't have a Koch brothers situation as in the US—powerful energy industry magnates pouring money into climate change denial, lobbying, and right-wing election campaigns.

GLOBAL GOVERNANCE

The consequences of climate change will be felt most at a local level, and yet what happens is enormously dependent on global governance. There is no global government, on this issue or any other, but there is a global regime or set of governance arrangements. At the core of this regime has been the UNFCCC, established in 1992 to negotiate global agreements. Under the auspices of the UNFCCC there is an annual two-week COP. Activity in and attention toward the global regime peaked in 1997 with the adoption of the Kyoto Protocol that committed developed (Annex One) states to greenhouse gas emissions reductions in a period extending to 2012, and again in 2009, the fifteenth COP in Copenhagen.

COP-15 was expected by some to produce a comprehensive global treaty for controlling emissions, but in practice produced a last-minute "Copenhagen Accord" negotiated by the US as represented by President Obama and China (the latter with some support from Brazil, India, and South Africa). This Accord specified a "guardrail" of a 2°C average global temperature increase over pre-industrial levels, and a set of pledges for different categories of countries which if secured would unfortunately only give a 50 percent chance of avoiding 3°C by 2100 (Rogelj et al., 2010). Also anything the US committed could not be taken too seriously, given the impossibility of getting any treaty ratified by the US Senate, or any effective action passed by the US Congress. The COP as a whole then voted to "take note" of this accord, rather than adopt it. It could not be adopted because the peculiar rules under which the UNFCCC operates mean that any country can object to any clause in any proposed agreement, and that objection must then be somehow dealt with. The result is often some convoluted language, "constructive ambiguities" that can satisfy the often contradictory concerns of all parties. Given that 194 countries participate in each COP, it is unsurprising that the process can move very slowly. Participants such as Saudi Arabia that only want to prevent action on emission reduction find it very easy to be obstructive.

The disappointing outcome at Copenhagen did not spell the end of the UNFCCC negotiations. Two years later in Durban in 2011 another last-minute agreement at least set out a timetable for further negotiations that would produce a binding agreement in 2015, and for the first time major developing countries such as China, Brazil, and (reluctantly) India agreed that their own future emissions trajectories were now seriously on the table. Compared to what preceded it the Durban agreement looked positive; compared to what needed to be done it looked minimal (see Light, 2011).

There are several ways the UNFCCC negotiations can be understood. The first is in terms of a collective effort by the world to solve a common problem. While this ethos was perhaps important in the 1990s, it gradually came to look less plausible than the idea that the negotiations were mainly about states or groups of states trying to protect their own material interests. So the G77 of developing countries plus China tried to shift all the responsibility for action to developed countries (deploying a historical justice argument of the kind we discussed in Chapter 5). The Association of Small Island States tried to push for a strong legally binding agreement to curb emissions. The "Umbrella Group" of the US, Canada, Australia, New Zealand, Russia, Ukraine, Japan, Norway, tries to get commitments from the large emerging emitters such as India and China. The EU sometimes looks progressive, in large part because it is moving forward on (if slowly and unevenly) ecologically modernizing its economy. Some developing countries see potential for massive financial benefit from REDD schemes through which others would pay them to maintain and/or create carbon sinks, such as forests or tree plantations. Oil exporters such as Saudi Arabia just try to be obstructive.

Another way of thinking about the negotiations is in terms of the discourses that are present or absent, strong or weak. A discourse of sustainable development or its close cousin ecological modernization is strong; but not strong enough to yield agreements that would effectively secure its principles. Climate marketization is increasingly powerful; the idea that global action should be pursued through emissions trading schemes or offsets (Paterson, 2011). A discourse of emphatic denial is completely absent in UNFCCC proceedings. A discourse of climate justice is strong, especially in terms of the established norm of "common but differentiated responsibilities"—though different parties have different views as to what justice demands, and the balance between what is common and what is differentiated across rich and poor countries (see Chapter 5).

These discourses are advanced not just within the negotiations, but also in the lively civil society activities that surround and sometimes influence the negotiations. NGOs, business associations, and activists of all sorts (except emphatic deniers) show up in large numbers—over 13,000 participants registered for entry to the conference venue in 2009 in Copenhagen. The most prominent NGO is the Climate Action Network, which also acts as a coordinator for other NGOs seeking action to combat climate change. Green radical discourses that are generally not present in the official negotiations (except in a few Latin American delegations, notably Bolivia) have a large presence in parallel civil society events—such as the Klimaforum organized in Copenhagen in 2009 that provided a venue several kilometers from the COP site for activists to gather and exchange ideas. All this civil society activity means a range of important concerns gets raised—if not always heard or acted upon within the negotiations, especially when NGO representatives are largely excluded in the final crucial days.

Participants in and observers of the global climate negotiations often see a defensible binding agreement as the proof of success or failure of global governance—hence the growing belief that the process since Kyoto has been a failure of global governance. Unfortunately, however, even the signing of such an agreement does not guarantee that national governments will comply with it. We have already noted that most of the Annex One signatories to the Kyoto Protocol fell short of their 2012 commitments. There is a still more insidious problem in that states are quite capable of negotiating the terms of their apparent compliance in ways that suit their own economic interests, but subvert the intent of the agreement. Stevenson (2012) demonstrates the variety of strategies that states follow here. Rich countries can offload production of the goods they consume onto poor countries, producing an illusory net reduction in their own emissions. They can buy carbon credits from poorer countries. They can purchase offsets from poorer countries that promise to grow trees—but in practice do no such thing (or even sell the same promises several times over).

REFORMING GLOBAL GOVERNANCE

Those exasperated at the mismatch between the glacial progress of the global governance of climate change and the urgency of the problem sometimes long for more effective centralized global authority. That is not realistic (though that doesn't stop the fear of such an authority driving the paranoia of American denialists). The closest thing the global system currently has to such an authority is the World Trade Organization—effective because it is set up to reinforce the core economic priority of states. The fact that global climate governance is not obviously beholden to any of the core priorities of states that we identified earlier in this chapter (security, revenue, economic growth, welfare) is at once a negative and a positive. It is negative because it may be hard to get states to comply with any global agreement if it counteracts their priorities. It is a positive because there may actually be more scope for innovation in global governance than there is at the level of the state. So now let us look at some reform proposals.

A wide-ranging reform agenda in the name of earth system governance is articulated in a 2012 *Science* article by Biermann et al. Initially staying close to existing institutions such as the United Nations, they call for the upgrading of organizations such as the United Nations Environment Program, the integration of environmental concerns into global economic regimes, and more in the way of equity and transparency. More expansively, they call for a "constitutional moment" such as that last seen after World War II. The problem is that there has never been such a moment of transformation of the fundamental institutions of the international system except in the wake of total war (as we pointed out in Chapter 1), and there is little reason to expect it to happen just because it is advocated by those who care deeply about global sustainability.

Exasperation leads observers such as Victor (2011) to suggest we should stop thinking of a comprehensive global regime for climate change, and instead pursue a more disaggregated strategy. He argues against the multilateralism of the UNFCCC and its fixation with a legally binding agreement with targets and timetables for emissions reduction. Victor believes a better way to proceed would be for the more enthusiastic proponents of action (such as most countries of the EU) to form "clubs"—coalitions of the willing. Members of these clubs would make (non-binding) promises to each other about the kinds of actions they will undertake, and develop institutions for technology development and transfer and carbon trading. The idea is that they would with time make it attractive for other countries to join the club in question, perhaps getting access to renewable technologies, or opportunities to participate in and profit from the club's emissions trading market. Unwilling countries could simply be ignored and would have no way to block deals.

The challenge for any disaggregated approach is that it is by no means clear whether or not the various initiatives sum to anything like an adequate and timely global response to limit emissions, temperature rise, and the impacts of climate change. One way to salvage reasonably comprehensive global agreement while getting round the problems caused by the need for the agreement of 194 negotiating parties is encapsulated by the idea of "minilateralism" (Naím, 2009). Minilateralism would require the participation of only a small number of large states. Naím suggests that 20 might be an appropriate number, because 20 states account for 75 percent of the world's greenhouse gas emissions. Victor (2011: 214) believes 12 would be enough. The number could be reduced still further if we see the EU as one negotiator. Any agreement among these states would have plenty of force, even were it not ratified by anyone else. Minilateral proposals generally include the US—but as we have seen, the US is a problematic partner in any global negotiations, and so probably unenthusiastic about any minilateral deal.

Attracted by the comparative feasibility of minilateralism but worried about a small number of countries seen to be ordering the world, Eckersely (2012) proposes "inclusive minilateralism" in the form of a global "Climate Council" composed of states that are the most responsible for the accumulated emissions in the atmosphere, most capable of acting, and representatives of the most vulnerable to the effects of climate change. Such a

Council might then be composed of the US, China, European Union, Russia, Japan, India, Brazil, South Korea, Mexico, plus representatives from the Association of Small Island States, the African Union, and the Least Developed Countries group. It is the inclusion of representatives of these last three "most vulnerable" categories of countries that distinguishes Eckersley's proposal from more conventional minilateralism, because the most responsible and most capable states are also the large emitters. The task of the Council would be to sort out long-term emissions reductions targets—essentially the shape of the "contraction and convergence" graph we introduced in Chapter 1 (see Figure 1.1).

For Eckersley, any agreement reached by this Climate Council would secure legitimacy through subsequent ratification by the existing 194-member COP. But that leaves out global civil society (NGOs, activists, media), which can both help hold global governance accountable and contribute to its legitimacy in the eyes of key actors around the world by conferring approval on processes and outcomes. We have already noted the vast amount of civil society activity surrounding the UN climate negotiations, but its connections to the formal processes are somewhat haphazard—as are connections between different parts of civil society, such as business-oriented and green radical forums. Perhaps that is the way it should be, if the role of civil society is to provide an arena for open communication, the generation of ideas, and critique of the status quo or of any agreements reached or on the table. But proposals have also been made for more systematic representation and engagement of the different kinds of interests, activists, and discourses present in global civil society (Biermann, 2011: 693). One could imagine any agreement being submitted to a body composed of civil society representatives for review and deliberation, or simply NGO representation along with states in an even more inclusive minilateralism. The danger would be that any such formalization of the role of civil society would entail cooptation and a diminution of its capacity for critique.

NETWORKED GOVERNANCE

Some new forms of climate governance have emerged that downplay or even bypass the sovereign authority of states as well as UNFCCC negotiations (though they can still develop relationships with states and the UNFCCC). Participants can include sub-national and local governments, agencies of national governments, corporations, NGO, and international organizations (see Hoffman, 2011 for a survey). Their common feature is the network form of organization, which can extend across national boundaries (but does not have to).

Climate governance networks come in several forms. The first is organized in order to administer international aid and development, technology transfer, and finance. The most prominent examples is the CDM organized under the Kyoto Protocol. This involves entrepreneurs creating projects to reduce emissions and install low-emission technology in developing countries; the funding comes from wealthy countries, for whom it is cost-effective to pay for clean projects in poorer countries, rather than reduce emissions at home. The CDM involves "private investors, carbon buyers and brokers, host governments, multilateral organizations, accredited independent verifiers and NGOs" (Bäckstrand, 2010: 89). The Clean Technology Fund is sponsored by the World Bank and is overseen by a Trust Fund Committee whose members come from eight donor and eight recipient countries, working cooperatively with businesses and NGOs. The Climate Technology Initiative's Private Financing Advisory Network relies, as its name suggests, much more on private sources of finance.

The second main kind of network organizes carbon trading. The Chicago Climate Exchange (which expired in 2010) was a private initiative that extended to corporations in several countries, including India and China, who could voluntarily buy and sell emissions permits. The trading of offsets too can be organized on a private basis; the Offset Quality Initiative provides oversight and quality control (Newell and Paterson, 2010: 123–5).

The third kind of network involves mutual commitment and information exchange (but no monetary exchange). City governments around the world have been active in developing ways to reduce their carbon footprints, and have developed networks such as the C40 Cities Climate Leadership Group, and ICLEI Cities for Climate Protection to share information, ideas, commitments, and encouragement (Bulkeley and Betsill, 2003). Networks involving both public sector and corporate actors are also emerging to explore or promote renewable energy, geo-engineering, and carbon capture and storage. The attractiveness of this third kind of network in particular is that it can move ahead based only on the participation of those willing and able; nobody need stand in their way. Many participants and supporters devote time, energy, and finance as a result of their disappointment with inaction at national and global levels. Their component institutions such as cities, businesses, and universities have changed the way they operate in order to reduce emissions. Their weakness is that currently they are quite marginal, and have no way of bringing more recalcitrant governments and polluters into line.

When networks do involve finance and trading, there is no clear evidence that they actually perform much better on climate issues than more conventional forms of governance, though that is partly due to the absence of obvious success indicators that would enable such comparisons (Pattberg, 2010: 153). The businesses that join them do so because they can see benefits when it comes to solidifying their own corporate social responsibility credentials, or can make money from emissions trading or offset schemes or new technologies. Gross polluters remain beyond the reach of all three kinds of networks. Existing governance networks give hints about what might be done, and might be effective, thus providing some resources for contemplation of more thoroughgoing governance reconstruction, to which we return in Chapter 8.

CONCLUSION

For different reasons, sovereign national governments and global governance have so far proven incapable of responding effectively to the range of challenges presented by climate change. While some countries do better than others, none does well enough. We have looked at ways that national governments might adjust their priorities, how the system of global governance might be reformed to the benefit of both its effectiveness and legitimacy, and the role of emerging forms of networked climate governance. Reconstructing governance for a climate-challenged society remains a major task, which we will address in the context of larger processes of transition in Chapter 8.

7

The Anthropocene

Climate change will challenge the human community in many ways for centuries to come. Human influence on the climate is now the primary driver of the shift to a less stable and more dynamic global environmental system—the Anthropocene. In this chapter we explore some profound implications of this new age. First, what we mean by "the environment" is now itself ever-changing, with human actions affecting the very makeup, functioning, and evolution of global and local ecosystems, pushing them in new directions that can be difficult to predict. Second, this new reality has consequences for the founding principles of environmental management, conservation, ecosystem restoration, and action on the environment in general. The use of the past as a baseline natural world to be restored or mimicked is no longer possible, and so the era of preservation as the basis of environmental management is over. Climate change is pushing ecological systems out of their Holocene comfort zone (the last 10,000 years of unusual climatic stability). Our conceptions of a "natural" world and how people relate to it will have to change as well. Scientific controversies, environmental politics, and ecological management begin to look very different as a result.

While most environmental scientists warn of the profound difficulties of navigating the Anthropocene, some technological optimists envisage a brave new future where humanity progresses through continued advances in biotechnology, information technology, and nanotechnology (Silver, 1997; Kurzweil, 2005). In this light, climate change and the transition to the Anthropocene are just a bump in the path of human progress. This kind of thinking extends to geo-engineering the planet to both avoid the worst of climate change and even push human development in new directions. While some climate scientists are beginning to explore the possibilities and consequences of geo-engineering, others are concerned that such bold action will exacerbate environmental uncertainties. These tensions among scientists represent competing visions of the degree to which governance informed by science can really understand and constructively guide Earth processes.

If humanity survives into the long run, there may be ways that the Anthropocene can be organized to provide for both ecosystem and human functioning. But that means taking responsibility for the reality of our eco-engineering and terra-forming selves, something we have not yet been able to do. This is the challenge of the Anthropocene.

ENTERING THE ANTHROPOCENE

Bill McKibben, now known primarily for his climate activism (he founded 350.org), originally made his name with the publication of *The End of Nature* (2006 [1989]). McKibben argued that rather than being "a species tossed about by larger forces—now we *are* those larger forces" (p. xviii), becoming perhaps the dominant force of change on the planet. McKibben's recognition was not altogether new. Historians had documented several millennia of human influence (Glacken, 1967). As long ago as 1864 George Perkins Marsh published *The Earth as Modified by Human Action*. Italian geologist Antonio Stoppani, citing Marsh, suggested an "anthropozoic" age of human forces on the Earth in 1873, and V. I. Vernadsky discussed human influence on evolution in 1926 (Crutzen, 2002; Steffen et al., 2011a).

Clearly, it is not new for human beings to alter, and sometimes destroy, ecosystems. Agriculture changed landscapes, hunting wiped out species. Humans have been serially depleting resources—taking them until collapse, and then moving to the next source—as long as we have roamed the landscape. Historical research on fisheries shows that we undermined ecological systems and species long before industrialization began (Holm et al., 2010). History shows that human societies are even capable of causing their own collapse by destroying resources on which they depend (Diamond, 2005), suggesting we should neither romanticize a past before large-scale burning of fossil fuels began nor be naïve optimists about the future. The Anthropocene represents a shift to an altogether bigger scale, on a more rapid basis. We are now talking not about impacts on various local systems, but on the entirety of the Earth system. Climate change is the great exemplifier of this new age of human influence. "Humankind will remain a major geological force for many millennia" (Steffen et al., 2007: 618).

Paul Crutzen, the Nobel Prize-winning chemist who began popularizing the idea of the Anthropocene in 2000 (Crutzen and Stoermer, 2000), argues that this era starts with the growth of CO_2 and methane concentration in the atmosphere in the mid-eighteenth century, revealed by polar ice samples (see also Crutzen, 2010). Will Steffen, joining Crutzen in a proposal to formalize this new age through recognition by geological societies, notes the same start, but argues that something changes in the mid-twentieth century. By 1945, atmospheric CO_2 concentration had risen by about 25 ppm above pre-industrial levels, "enough to surpass the upper limit of natural variation through the Holocene, and thus provide the first indisputable evidence that human activities were affecting the environment at the global scale" (Steffen et al., 2007: 617). Since then, CO_2 concentrations (and many other measures of environmental impact) have risen at an increasing rate, leading Steffen to label this very recent past "the great acceleration" of human impacts on the global ecological system (Steffen et al., 2007: 616–17).

In addition to changing average temperature and precipitation patterns, climate change and other drivers of the Anthropocene (such as land use practices and pollution) are having a major impact on biodiversity. Life on Earth is in the midst of the sixth great extinction, with species disappearing at 100–1000 times the expected or background level. We are no longer simply serial extinguishers of particular species like mastodons and passenger pigeons in particular places. Rather, human activity is decimating the biodiversity of the planet on a broad scale.

The idea of a new Anthropocene era dramatizes not only human impacts on global systems, but also the fact that those impacts are pushing the Earth system out of the relatively stable past 10,000 years of the Holocene. Human beings have been very well-served by a stability that enabled settlement, agriculture, and civilizations to flourish. Such a stretch of time is only a fraction of human life on the planet, but it encompasses all of recorded human history. That nourishing and stable Holocene, scientists argue, could have continued for another 10,000 to 20,000 years, if not for the interference of human beings in global systems, including climate. We see an "unintended experiment of humankind on its own life support system" (Steffen et al., 2007: 614) of which we have now become a part. As the *Economist* (2011a) puts it, human beings "are not just spreading over the planet, but are changing the way it works." We can no longer look out at nature without looking back at ourselves. Once we could move on from depletion and waste, establishing new communities in new places. In the Anthropocene there is no longer any frontier, no place to which we can escape.

How then do we manage in this brave new world? We have many ways to mitigate and adapt, as outlined in Chapter 4. Our current understanding of the global system informs us that change in our behavior is needed, but is our knowledge good enough for the task? Given humankind's existing problems in adjusting behavior to emerging environmental knowledge—or even taking that knowledge seriously—can we meet the challenge of the Anthropocene?

BOUNDARIES AND LIMITS

A group of prominent scientists led by Johan Rockström, writing in the influential journal *Nature*, offer one way of thinking about how to proceed in the Anthropocene. These scientists argue that there are a number of "planetary boundaries that must not be transgressed" (Rockström et al., 2009: 472). These boundaries define the "safe operating space for humanity with respect to the Earth system." The idea is to limit our influence on the planet's key biophysical systems, to keep us from crossing "thresholds" or "boundaries" that would "push the Earth outside the stable environmental state of the Holocene, with consequences that are detrimental or even catastrophic for large parts of the world" (Rockström et al., 2009: 472). We should manage the character and levels of human activity to stay within the boundaries.

Rockström et al. identify nine processes, and a boundary for each. They are: climate change; rate of biodiversity loss (terrestrial and marine); interference with the nitrogen and phosphorus cycles; stratospheric ozone depletion; ocean acidification; global freshwater use; change in land use; chemical pollution; and atmospheric aerosol loading. They specify the pre-industrial value of each indicator, its current status, and the proposed boundary. The precise boundaries are consensus judgment calls (which could be adjusted later), rather than determined directly by any scientific procedure. For climate change, the boundary is 350 ppm of CO_2 in the atmosphere, so (along with those for the nitrogen cycle and biodiversity loss) it has already been transgressed—at the time of writing we are just surpassing 400 ppm.

A famous precursor to planetary boundaries appeared in 1972, the Club of Rome's *Limits to Growth* (Meadows et al., 1972), which argued that exponential growth in population and economy would eventually hit limits. The limits in question were defined by finite stocks of natural resources and the capacity of the biosphere to assimilate pollution. For Meadows et al., humanity had several decades to develop a steady state economy that would avoid overshoot of limits and subsequent social collapse. Similar metaphors of limits transgressed, boundaries overstepped, thresholds crossed, capacity reached, and overshoot causing potentially devastating feedbacks, join Meadows et al. and Rockström et al.; though Rockström et al. have little to say about finite stocks of natural resources, and avoid the word "limits."

However justifiable it may be in ecological terms, dominant actors in political and economic systems do not like such talk of boundaries, as any environmental advocate can attest. Of course, all developed countries and most developing countries now control air and water pollution, pesticide use, and other environmental harms, informed by scientific arguments that some level of effect is acceptable but going beyond this level is not. Larger arguments about limits and boundaries are a much greater threat to the political and economic status quo. Those committed to this status quo often paint environmentalists as anti-growth in order to dismiss them. As two of the current authors have written previously (Dryzek et al., 2003), states have an imperative *for* growth, and environmental discourse that is couched against that imperative has rather consistently failed in the last few decades. Rockström and his colleagues represent mainstream ecological discourse, but their metaphors have so far proven not to work in politics. Just as the *Limits to Growth* argument of the 1970s had little or no impact on the policies of governments, so arguments for planetary boundaries to human activity have thus far had little effect in modifying responses to climate change at any national let alone global level. At the landmark 2012 United Nations Conference on Sustainable Development in Rio de Janeiro, planetary boundaries were mentioned in the draft conference declaration—but deleted from the final version as too controversial. While reasonable from an environmental science perspective, planetary boundaries as a guiding discourse for the Anthropocene remain problematic politically.

NO GOING BACK: THE PAST AS GUIDE POST, NOT GOAL POST

Setting aside the political problem, there is also a conceptual problem in linking planetary boundaries to the Anthropocene. The effort of environmental scientists to determine planetary boundaries is based on the assumption that we can with the right actions return to them safely after we have exceeded them. In this light, planetary boundaries are not guidelines for living in the Anthropocene but for avoiding it. And yet the whole idea of the Anthropocene is that going back to a stable Holocene is not possible. The world is changing, and that has implications for the level at which any boundaries are set. Boundaries as currently presented look static, while the Anthropocene is dynamic—and threatens to render our knowledge of the history of past environments somewhat irrelevant to future practice.

Management of environmental systems has traditionally depended on such knowledge: the historical rate of snowmelt and flow of rivers, the range and migrations of various flora and fauna, the history of take, size, and species in fisheries. Of course, the past is all that we will ever have from which to learn, and environmental management has long taken the past as a standard around which to design conservation and restoration. Yet our move into the Anthropocene, out of the relative stability of the Holocene, undermines this knowledge basis of management by snatching away the trusty compass provided by the past.

This connection to history and reference to past, stable conditions is deeply embedded in standard terms like "preservation," "conservation" biology, "restoration" ecology. But while the past can no longer be our model, we can still look back to understand the nature of "healthy" and more resilient ecosystems, and so guide our response to a rapidly changing and unstable Anthropocene. So the past is relevant, but in new ways that environmental scientists are only beginning to understand. Transferring this new understanding to the public and policy makers will take time, especially if some stakeholders (for example, resource management bureaucracies) cling to old understandings.

In the Anthropocene, the very idea of environmental preservation is compromised, as everything we call wilderness is affected by human actions—even if no human physically enters it. Cronon (1995) famously made the argument that "'nature' is a human idea" and wilderness "quite profoundly a human creation." While attacked by purist wilderness advocates, the argument was prescient: the Anthropocene makes the idea of human-made nature all too plausible. The wilderness we have imagined becomes less relevant as a managerial guide.

In addition to problematizing the idea of wilderness, Cronon was attempting to point out that environmentalists and environmental managers have often used notions of "nature" or "wilderness" as uncontested categories that no right-thinking person should resist. As soon as we label something natural, he argued, we attach a powerful implication: that any change from this state means damage and degradation (p. 20). Cronon's reflective rant against the idea of wilderness was not an argument against care for the natural world. Rather, it was a call for us to question over-simplification of the "natural" wilderness idea, to recognize the variety of possible meanings, processes, and uses of the natural world, and to engender discussion across cultures, perceptions, and conceptions of the human/non-human interface.

The field of ecological restoration serves as an example here. Restoration ecology originally defined itself in terms of moving ecological systems back to indigenous, historic ecosystem conditions, before damage began. The practice can be traced back to Aldo Leopold's work in the 1930s in restoring Wisconsin grasslands. Remove historical benchmarks, and we not only undermine restoration ecology's *raison d'être*, but also illustrate the limits of our ability to rehabilitate nature as changed by the Anthropocene.

Restoration ecologists have begun to think about what it means when the past becomes less relevant, or important but in significantly new ways. Eric Higgs (2012) has recently reflected on the future of restoration ecology in an age of climate change. While history cannot provide fixed reference points, Higgs still sees some form of "historical fidelity" as a virtue appropriate to a rapidly changing nature. In 2002 the Society for Ecological Restoration (2004) defined restoration as "the process of assisting in the recovery of an ecosystem that has been damaged, degraded, or destroyed." In line with this definition, Higgs argues that the practice is no longer about restoration of conditions present at some specific time, but rather about removing sources of degradation, and restoring beneficial processes. Historicity in this light is about the way that ecosystems evolve, longstanding cultural relationships with landscapes, and human responses to shifting ecological conditions. Sandler (2012) distinguishes between strong and weak historicity—strong uses the past to set restoration goals, and the weaker form (as advocated by Higgs) uses historicity to support ecosystem recovery under new conditions. As Throop (2012: 49–50) interprets this new development, "restorationists should seek to move an ecosystem forward toward a structure that reflects what people value about the past, while responding to changing situations."

This transition from thinking of ecosystems in static terms to dynamic terms reflects the same transition that Earth system modelers have made. It is well underway among theoretical ecologists, but not yet a part of common public understanding. It has had trouble reaching even applied biologists who practice restoration ecology and conservation biology, let alone the broader public and policy makers. And as with any scientific understanding, its acceptance is now impeded by declining deference to scientists, at least in the US.

TAKING THE REINS OF A SPOOKED HORSE

In the Anthropocene, human beings have become major agents of environmental change—but have yet to come to grips with all that implies. As Steffen and colleagues put it (Steffen et al., 2011b: 749), "We are the first generation with the knowledge of how our activities influence the Earth System, and thus the first generation with the power and the responsibility to change our relationship with the planet." But what does it mean to exercise this power responsibly? Without necessarily intending to do so, we have taken the reins of the global environmental system that sustains us and all other entities on the planet. But so far we have shown little in the way of good horsemanship.

Responsible exercise of this power cannot mean business as usual as advocated, for example, by organized climate change deniers, or taken for granted by those enraptured by short term concerns about the economy. Nor can it mean simply trying to restore ecosystems to some historically pristine state, as we have just argued. Rather, responsible action must mean taking on collective responsibility for what happens to the processes of the Earth system (which can no longer be described as natural) in such a way as to promote their sustainable and sustaining character.

If business as usual and return to the past are ruled out as basic guides for responsible action, that still leaves a range of possibilities. We will now address three prominent ones. The first embraces the Anthropocene by expanding human power to engineer the planet—or even engineer ourselves. The second sees the Anthropocene as a condition to be administered. The third seeks a more co-evolutionary relationship encompassing humans and ecosystems.

THE ANTHROPOCENE ON STEROIDS: GEO-ENGINEERING AND HUMAN ENGINEERING

If so far the consequences of human-induced changes to earth systems look negative, one response is to change the content of human intervention to produce positive consequences. This could be done in a big way through engineering the environment. Keith (2000) was one of the first to advocate the idea—publicly, anyway—in a work exploring the prospects for geo-engineering the climate system to avoid the worst impacts of climate change. Paul Crutzen (2002, 2006) came out in support of at least research on geo-engineering, in case we become desperate—if we prove unable to achieve sustainability any other way. Geo-engineering would be the last line of defense against climate catastrophe.

Several geo-engineering technologies have been proposed, though none is proven. Seeding the oceans with iron or other nutrients would stimulate marine plant growth and so take-up of CO_2. More radically, it is possible to imagine the development of carbon absorption machines to suck CO_2 out of the atmosphere. It is also conceivable that genetic engineering of plants could increase their ability to absorb and store carbon. Currently though the most popular idea is that suggested by Crutzen (2006), which would involve injecting sulfate aerosols into the atmosphere to block solar radiation and, so, the heating of oceans, ice caps, and land. One reason for the popularity of this technology is that we already have examples of it working in the way we would want—the well-known cooling effect of volcanic eruptions. Large eruptions such as that of Pinatubo in the Philippines in 1991 yield measurable decreases in average global temperature that can last two to three years—in Pinatubo's case, over 0.5°C. Solar radiation could also be blocked more effectively by clouds that were thicker and whiter, and this could conceivably be done by machines that injected water vapor or droplets into the atmosphere from the oceans.

Geo-engineering advocates believe it would cost much less than shifting human energy production away from fossil fuels. An article in *The Economist* (2011a) proposed the use of geo-engineering to both manage global temperatures and artificially to recreate benign Holocene conditions. "Embracing the Anthropocene," *The Economist* argues, would mean shaping desired environmental conditions, rather than "retreat onto a low-impact path" indicated by the presence of boundaries.

Geo-engineering sees the Anthropocene as presenting a technical problem, which can be solved through technological means. As such, it draws on a Promethean tradition of long standing.

A number of technical and ethical issues should give us pause here. Not much is known concerning whether or not the various technologies would work as predicted, and some of them remain distant dreams. Spraying sulfate aerosols into the atmosphere might reflect sunlight and so keep temperatures down, but it would do nothing about the growing saturation of the atmosphere with greenhouse gases under that sulfate umbrella. If the machines were to shut down for any reason, we would very quickly feel the catastrophic effects of a major immediate rise in average global temperatures. The machines must never stop, demanding a level of continuity in human institutions on a scale never yet seen in our history.

Geo-engineering may also bring "unintended and unanticipated side effects that could have severe consequences" (Steffen et al., 2007: 620). We do not know who or what will turn out to be hurt. Geo-engineering is an experiment on the global population, and as we saw in our discussion of adaptation in Chapter 4, some are better placed than others to cope with rapid and unpredictable environmental change of any sort. Philosopher Steve Gardiner points out that climate change establishes a moral hazard; we should be wary of arguments and practices that appear to diminish our moral responsibilities (Gardiner, 2011: 345). The lure of geo-engineering frees us from responsibility for the behavior that creates the problem of climate change in the first place.

Geo-engineering may appear radical, but a more recent proposal takes control over nature still further—and yet in a way closer to home. Liao et al. (2012) suggest we think about engineering not the planet, but ourselves. Human engineering, they argue, lies beyond both behavioral change (such as that induced by carbon taxation) and geo-engineering. "The biomedical modification of humans to make them better at mitigating climate change" (p. 207) could include inducing meat intolerance, making humans smaller (to shrink the carbon footprint along with our actual one), "lowering birth-rates through cognitive enhancement" (p. 209), and "pharmacological enhancement of altruism and empathy" (p. 210). We human beings would manipulate our own evolutionary path in order to better fit with the environment we have created. The authors argue that human engineering is much less risky than geo-engineering—and that it complements existing behavioral and market-based approaches.

Liao et al. admit that theirs is a speculative exercise; they "wish to highlight that examining intuitively absurd or apparently drastic ideas can be an important learning experience" (pp. 216–217). Such talk may provide plenty of ammunition for hardline climate deniers already convinced that concern for the climate harbors a sinister totalitarian agenda. And yet, Liao et al. suggest that the idea might become less controversial in future—and seen not as a limitation of freedom, but its expression. For just as we are inadvertently affecting global environmental systems, so we are already engineering our own bodies and offspring (inadvertently toward maladaptations such as obesity, but also purposefully against illnesses). In this light, if genetic modification is contemplated for health reasons, why not also for environmental reasons? As wild as it may sound, negotiating the Anthropocene could then cover design of humans as well as environments.

MANAGING THE ANTHROPOCENE

Geo-engineering represents extreme environmental management, human engineering a yet more radical form. But ecological management in the Anthropocene could be a more multifaceted affair of organized collective responsibility for planetary life-support systems. It might be possible to start with some general principles of good environmental management: looking forward rather than back, understanding and working with natural systems, planning for complexity, and being willing to reflect on knowledge shortcomings as we take control over the trajectory of Earth systems. In a rapidly changing world, we would need to keep updating our science and use this knowledge to devise new rules and incentives to guide actions. In this light, it is possible to imagine taking the reins of the spooked horse through good administrative practice guided by science and informed by a concern with the wellbeing of whole systems.

Such practice could build upon existing approaches to ecosystem management (such as adaptive management), though with unapologetic interventionism in order to reconcile and regenerate human and natural processes in order to provide life-support services (Higgs, 2012: 98). Climate change, argues Throop (2012: 48), "may liberate ecosystem managers from the past and enable them to create novel ecosystems that maximize utility"— it might help us abandon " 'purist' ideas of the way ecosystems should be, and instead emphasize protection of the ecosystem services they provide for us." Ecosystem services here refer to goods provided to humans: renewable resources, moderation of environmental conditions, recycling of pollutants, aesthetic refuge, recreational opportunities, and so forth. Management would feature protection and redesign of human-ecological systems in order to maximize the flow of services they provide so as to enable societies to flourish.

One of the fields that has begun to examine such an approach is urban design and planning—the growth of green architecture and the development of plans for green cities embody the idea of designing everyday human life around the processes of the natural world. From a concern with minimizing urban heat islands by creating roof gardens and extensive parkland, to local renewable energy production, to sustainable food production (and even high-rise farming), ideas once considered niche or even fantasy are becoming mainstream.

However, larger-scale development of bio- or eco-mimicking sites (Benyus, 2002) is easier said then done. And urban eco-design leaves untouched bigger questions concerning the management of ecosystems at regional and global levels. Broome (2012) argues that the management of such systems is possible once our large brains have identified the danger; we've responded to crises in the past. But as we stressed at the beginning of this book, climate change presents an altogether different kind of challenge. While it is possible to imagine management that comprehends the science and applies it to the regulation of regional and global ecological processes, it is also possible to imagine catastrophic cascades of unintended consequences as managers attempt to grasp that which is beyond their reach.

How then to ensure that management is up to the task, especially in light of the fact that currently we have no good examples of global ecological administration? Braithwaite (2010) argues for a managerial method that explores a variety of potential futures, rather than adopting a single idea or pathway—what he calls "scenario framing." Such an approach allows for a flexibility and adaptability in management. However, there may be a problem embedded in the very idea of top-down management, however flexible and adaptive it proclaims itself. Who is in control? Who coordinates? How can central administrators guard against the well-known tendency to oversimplify systems that are under their control? The logic of management implies central control of a type that is currently lacking at any level above that of the nation-state (and even there it can be absent). It is all very well to say as Lovejoy (2012) put it in a *New York Times* op-ed, "The moment has come to realize that this planet which brought us into existence must be managed as the biophysical system that it is. It is time to get our hands on the steering wheel." Unfortunately, there is currently no steering wheel. Who is going to install it? Invoking benign central control of complex systems is easy to do in the abstract, but devilish once anyone tries actually to put it into practice. Again we can look back on the *Limits to Growth* experience of the 1970s: pleas for more effective centralized management of the environment and natural resources had no impact on the structure and organization of government anywhere.

While it is possible to load all kinds of good qualities into a kind of management that would work for the Anthropocene, there may be a basic problem with the metaphor of "management." That metaphor moves us too quickly to the idea that we can organize a hierarchical system to assert control over and thenceforth administer human-ecological systems. While not necessarily entailing the degree of hubris associated with geo-engineering and human engineering, it assumes a capacity that is unproven. It also diverts attention from the fact that what is really needed is to begin with a thoroughgoing understanding of what it means to be complicit in the dark side of the Anthropocene and its degradation of global systems. For better or for worse we are *in* those systems—and have caused substantial damage as a result. Would it be any better if we were then to assert ourselves *above* those systems, as the "management" metaphor implies? Ecosystems are examples of a more spontaneous kind of order that arises from the bottom up. It is by no means obvious that the best way to enable them to provide planetary life support more effectively would involve completely changing the way we know they work in the name of management.

Yet the real difficulty with a managerial response to the Anthropocene may not be that it goes too far, but rather that it does not go far enough. As we pointed out in Chapter 1, responding to the challenge of climate change is going to require many linked things, not one big thing. Management is just one thing that needs to change, rather than something with a moderate record that should be picked out and placed above all else in the hope it can somehow transcend its past level of performance.

CO-EVOLUTION AND ECOLOGICAL RATIONALITY

It is possible to imagine a place for humanity that involves co-evolution with, rather than control over, Earth systems in the Anthropocene (on the basic idea of co-evolutionary development, see Norgaard, 1988). The first clue may be found in, of all places, *The Economist* (2011b), which amid its hubris also speaks of "finding ways to apply human muscle with the grain of nature, rather than against it...." Taking this literally would involve working *with* the processes of ecosystems, rather than seeking to dominate them. Figuring out how best to work with ecosystems would in turn require heightened critical reflexivity about our ecological selves. As Thompson and Bendik-Keymer (2012: 7) put it, this would mean "adjusting our conception of who we are to appropriately fit the new global context." We would then live with constant awareness of the environmental systems in which human life is immersed, and how we contribute to reworking the evolutionary path of the planet and our selves. Our success in navigating the Anthropocene would depend on our ability to understand environmental systems and reflect on our own place in them, and how we should act in consequence. We can expect Earth systems to keep changing in ways that surprise us, and so need to create individual and social capacities to both listen and respond to these changes.

Another way of putting it is that contemporary human beings and human social systems need to attain a degree of ecological rationality that has hitherto eluded them. Ecological rationality means sensitivity to feedback on the condition of ecological systems, recognizing the limits of individual knowledge and accepting the corresponding need to work collectively with proximate and distant others, prioritizing collective goods over individual material consumption, joining endeavors to repair damaged social-ecological systems (Dryzek, 1987). Ecological rationality also means reworking dysfunctional human systems, be they markets that operate as though nature has value only as an input to production, administrative hierarchies that cannot recognize local and temporal variation in problem conditions, or international regimes that seem incapable of delivering decisive outcomes of any sort.

Just to list these elements of ecological rationality points to social, political, and economic worlds very different to those we have now. However well our current individual sensibilities and social arrangements could cope with the Holocene, they are not adequate when it comes to the Anthropocene. Yet to point for the need for this kind of transition is not just an exercise in ungrounded wishful thinking. As we will see in the next chapter, intimations of these new kinds of individual sensibilities and social arrangements can be found in new thinking about resilience, in new forms of governance, economic organization, and technology, in new sorts of collective deliberation, and in new kinds of ethics and movements. Together, these practices provide the ingredients for a more de-centered and pluralistic but nevertheless coordinated response to the challenge of climate change and the Anthropocene.

8

Transition, Resilience, and Reconstruction

In the long run, responding to the challenge of climate change is going to require a de-carbonized economy with different energy systems and reconceptualized social-ecological relationships. Given the magnitude of the task, quite how to move in this direction remains a matter of some contention. In this chapter we examine several potentially complementary approaches, though nobody has yet identified *the* key to transition, and we do not resolve all the big questions about how to proceed. This coverage of a variety of innovations also makes sense in light of the failure to date of established centers of power—be they the global economic system, UNFCCC negotiations, or national governments—to craft effective responses. We group innovations under social resilience, new thinking about economics, new movements that embody this thinking, and new governance. Some developments are incremental, some more radical. These developments give us something more upbeat to end on, showing that transition is already being lived, if only, so far, on the margins of societies.

FROM ADAPTATION TO RESILIENCE

In Chapter 4 we looked at the importance of adaptation, especially in light of current global failure to advance on the mitigation front. Adger et al. (2011a: 757) point out that "adaptive responses are not equal in terms of the sustainability of resource use, energy intensity, reduction of vulnerability, or in the distribution of their benefits." While the international community has had difficulty in coming to terms with adaptation (even by 2011 the Green Climate Fund adopted at the Durban meeting of UNFCCC covered adaptation but omitted any clear definition of the term), it does hold significant potential. Adaptation does not have to be defeatist, and indeed can be linked to empowerment—we adapt by recognizing "that human beings can protect themselves from damage by living harmoniously with their atmospheric environment" and reduce their vulnerability to climate change (Burton, 1994: 15). In this section we consider how adaptation might contribute to transition.

On the academic side, adaptation to environmental change is increasingly understood in terms of the concept of resilience. According to the multinational Resilience Alliance (2013), "A resilient ecosystem can withstand shocks and rebuild itself when necessary. Resilience in social systems has the added capacity of humans to anticipate and plan for the future." Resilience means responding to changed conditions by reorganizing while retaining "essentially the same function, structure, identity, and feedbacks" (Folke et al., 2011), and so allows that shocks can provide opportunities for restructuring. Societies that manage for resilience ought to sustain "desirable pathways for development in changing environments where the future is unpredictable and surprise is likely" (Folke, 2006: 254). Resilience can be linked to governance that can respond to change and surprise—"away from policies that aspire to control change in systems assumed to be stable, towards managing capacity of social-ecological systems to cope with, adapt to and shape change" (Folke, 2006: 254).

Resilience can be thought of as a property of ecological systems, of social systems, and—crucially—social-ecological systems. Clearly many of our political systems are not resilient in the face of ecological shocks, be they liberal democracies or dictatorships. New Orleans and its environs demonstrated a lack of resilience when hit by Hurricane Katrina in 2005. Pointing to positive examples of resilience is harder, though perhaps resilience is best seen as an aspiration and a criterion for everything from individuals to the global system, rather than something already achieved by anyone or anything.

Ideally, then, resilience means flexibility, creative adaptation, and constructive relationships with ecological processes. Many of its proponents, for example John Barry (2012: 64), treat resilience as a prime "sustainability virtue," and see only positives. Barry believes resilient communities would feature high levels of solidarity, low levels of socioeconomic inequality, and empowered citizens (p. 27). There are, however, limitations. Resilience could also describe the economic and financial systems (including derivative markets) that brought the world to the edge of ruin in the global financial crisis in 2008, then managed to bounce back in a form that still imposed risks on everyone else. Markets also show resilience when confronted with climate change, with carbon trading and offsets especially popular, if not obviously effective (see Chapter 4). Systems can be resilient in themselves while destructive in their broader human and ecological consequences. Some critics even argue that resilience might displace the idea of poverty reduction, and offer a justification to insist the poor and powerless simply be resilient in coping with an unjust world, while not resisting it or trying to change it (Reid, 2012: 74). Resilience, then, may deflect attention away from the root causes of vulnerability to climate change (Cannon and Müller-Mahn, 2010: 623). Catney and Doyle (2011: 190) take this fear to an extreme:

> This current green welfare policy push to create *resilient communities*, is conservative at its core. It invents a green sphere which is premised upon the notion that we need to return society back to a steady-state, *before* the great ecological disruption. Indeed, the very definition of resilience is based upon the abilities of ecological communities to restore natural order as quickly as possible after disturbance.

Resilience might then go the same way as sustainable development: a promising notion co-opted to serve the status quo.

These criticisms and worries are warnings that should be taken seriously, but responding to them is straightforward. Adger et al. (2011b: 705) stress we must always ask: "resilience of what, and to what?" So when a system becomes untenable, its resilience is not desirable. The system may then need to be more thoroughly transformed—for example, the global political economy certainly needs de-carbonizing. Increasingly, thinking about resilience has taken on many of the same themes as recent work on development out of poverty; both focus on alleviating the vulnerability of individuals and communities when it comes to a range of basic capabilities and needs. We noted in Chapter 5 that the capabilities approach to development, embodied in the Human Development Index and Millennium Development Goals, can be applied in a straightforward way to address adaptation, development out of poverty, and climate justice by improving the ability of individuals and communities not just to cope with threats, but to progress on a variety of fronts (see also Schlosberg, 2012). Moreover, as one of us argued long ago, resilience only makes sense as one criterion among several for the assessment of social-ecological systems (Dryzek, 1987). In practice, we do see the concept of resilience inspiring a range of constructive and dynamic responses to vulnerability (not just desperate attempts to restore the status quo), and a host of localized practices of adaptation and development.

NEW ECONOMY

Resilience is then more than just a defensive measure, and does not rule out thinking more forcefully about restructuring economic and political processes for a climate-challenged society. We turn first to economics.

Economic systems change—locally, nationally, and globally. Thousands of years ago, the development of agriculture transformed how people lived much as we now see new information technologies transforming work, markets, and life. Nature can play a hand in transformation. Civilizations have crashed after depleting their resources. Ideas also matter. The clash between intellectual visions of people coordinated through ideal free markets versus ideal scientifically designed and implemented socialism underlay much of twentieth-century political and economic history. And as economies change, people's understandings of what constitutes a good life also change.

Can economies change fast enough, and in the right ways, to avert the worst of climate change? During the twentieth century, Western economies quickly transformed from market-driven to government-led during the Great Depression and World War II, then to mixed economies after 1945, before market fundamentalism returned in the 1980s, with a subsequent drive to globalization. The Russian and Chinese economies, once thought locked in by Marxist ideology, also turned to markets. Capitalism now has the upper hand globally. Could it be transformed into "climate capitalism" where emissions trading would work well, offsets would be effective, poorer countries would finance their clean development by the sale of carbon credits to businesses in wealthier countries? All this would be supported and regulated effectively by governments and international organizations, backed by financial institutions, and populated by businesses making profits out of renewable energy and the transition to a low-carbon economy forming a powerful political constituency to fight for de-carbonization (Newell and Paterson, 2010: 161–6). Pacala and Socolow (2004) argue we can de-carbonize the economy just by fully implementing numerous *existing* technologies, each addressing a small percentage of the need. The promise would be that of ecological modernization (see Chapter 6), simultaneously to grow the economy and protect the environment. Successful climate capitalism would have to find ways to suppress its darker side where marketization produces a playground for creative accountants and clever manipulators of complex financial instruments, yielding only fictional emissions reductions (such as "hot air" of the sort that post-Soviet countries could sell simply because their heavy industries collapsed in the 1990s) and unverifiable claims about offsets (Newell and Paterson, 2010: 166–9).

Defenders of climate capitalism would say that deep structural economic change is a distant prospect, and that if we really want effective action in any foreseeable and feasible future, we just have to work with capitalism, rather than rail against it. But rather than gamble on a climate capitalism that requires continued growth (and hope it is not too damaging), we might ask more radically whether the kind of growth to which existing societies, economies, and governments all seem to be addicted is actually necessary. Now, the idea of economic growth currently dominates, to the degree it is equated with human progress, not just the proper condition of a healthy economy. Thus, heretofore, calls for an economy without growth to stay within resource and environmental constraints have been chided as unrealistic, and gone unheeded.

What, however, is growing? Not happiness for the middle class and rich who have reaped much of the material rewards of growth. The literature on happiness indicates that, after a person has reached a modest level of income, more income only increases happiness temporarily, like the fleeting thrill of a major purchase. Real happiness comes through freedom, health, security, and good relationships. Has freedom increased or are people trapped in a system of markets and technological change that has colonized our lives, increasing our choices between supposedly different types of shampoo, while reducing our ability to live exciting, meaningful lives? The data on more hours spent working, growing obesity, declining psychological health, higher rates of incarceration, and suicides in the US, once the economy to emulate, raise serious questions even as gross national product grows.

In the 1960s and 1970s, a few economists, notably E. J. Mishan in the UK and Herman Daly in the US, began to question whether the gains from economic growth outweighed the environmental and social costs. At about the same time, other economists began to question whether growth made people happier (Easterlin, 1973). More recent works both document and fuel increasing unease (Bok, 2010; Easterlin et al., 2010).

In 2008 French President Sarkozy established a Commission on the Measurement of Economic Performance and Social Progress headed by some of the world's leading economists: Amartya Sen, Joseph Stiglitz, and Jean-Paul Fitoussi (and including five Nobel laureates in economics). At its establishment, Sarkozy (cited in Stiglitz et al., 2009: v) stated:

> If we do not want our future and the future of our children and grandchildren to be riddled with financial, economic, social and environmental disasters, which are ultimately human disasters, we must change the way we live, consume, and produce. We must change the criteria governing our social organizations and our public policies.

The Commission concluded with a report (Stiglitz et al., 2009) criticizing the adequacy of existing measures for managing economies, showing how governments could guide their economies through reference to measures of human ends served rather than simply by whether economic activity has increased.

In 2000, the UK established a Sustainable Development Commission that soon turned to consumption as a central problem. That Commission eventually yielded Tim Jackson's landmark 2009 book on *Prosperity without Growth*. Jackson (2009) identified a number of elements of a new economy, including an emphasis on "low-carbon, labour-intensive activities and sectors" (p. 176), investment in low-carbon economic infrastructure, changing national accounts to remove the obsession with gross domestic product, using human capabilities satisfied as a better measure, reducing work hours, promoting material equality, restoring social capital, and undermining consumerist culture.

Nongovernmental organizations including the New Economics Foundation in the UK and the New Economics Institute and the Institute for New Economic Thinking in the US document similar possibilities for creating economies that serve people and the environment, but not growth at any cost. James Gustave Speth is regarded as the voice of establishment environmentalism in the US: he founded and led the World Resources Institute, directed the United Nations Development Program, and was Dean of the Yale School of Forestry and Environmental Studies. Eventually he too became convinced that the underlying nature of modern economies must change to improve the course of humanity and the environment (Speth, 2012).

Economic growth has long been the core imperative of governments in market economies. But economies can change in the future as they have in the past, and life in the Anthropocene will bring new ideas, new technologies, new possibilities, and new challenges. The new economic thinking we have surveyed envisages economies that do not grow in conventional and unlimited material terms, while serving ourselves and our heirs just as well, if not better. But is there any relevant existing practice that points in the same direction?

LIVING THE NEW ECONOMY: NEW MATERIALISMS

We can discern hints of what a transformed and resilient economy might look like in an array of efforts seeking deeper change that encompasses new technological relations with the natural world, new economic and cultural institutions, and expressions of justice and empowerment. Several decades of environmental discourse stressing sustainability, resilience, post-materialism, and justice have informed activist understandings but at the same time have left palpable frustration with the limits of traditional kinds of action to deliver. Individual consumers might (if they can afford it) shop for "greener" goods at Whole Foods or Patagonia—but that hardly changes the dominant economy. Lobbying and pressure group politics has had its successes—but has not changed the core logic of the political economy. When growing food entails massive inputs of carbon-based energy, pollutes land and water with petrochemical additives, and alienates us from our local world and its seasons, sustainability and resilience seem distant dreams. In this light, it makes sense to question how we relate to the natural world, and how we provide our basic needs. Sustainability can be sought not only in individual purchases or expressed political interests, but also through more thoroughly reconstructive practices.

Movements for a new and sustainable materialism seek to change relationships with "resources" and nonhuman nature in everyday life. Environmental concerns are sometimes interpreted as emerging only after basic needs are met; Inglehart (1997) has long argued they are "post-material." Unmet post-material concerns can however inspire direct involvement in the flows of materials that meet basic needs, especially food and energy. New materialist movements want to replace unsustainable institutions, and forge alternative productive and sustainable practices at the local and regional level. The concern is with the flow of food, matter, energy—the stuff of everyday life—from the natural world, through productive processes, into and through our bodies, and back into the nonhuman realm. The point is to reconfigure flows that currently undermine integrity and resilience. So rather than simply support better-quality food or post-carbon energy, or use standard political tools of environmental interest groups to change policy, these citizens and community groups are themselves developing, participating in, and enjoying the products of new food and energy systems.

One example here is the transition town movement, which started in Ireland (in Kinsale) and then the UK (its spiritual home is the town of Totnes in Devon) and grew into a larger network. The movement began in response to the perceived failure of international, national, and regional government to confront peak oil and climate change. It intends to put resilience into practice by lowering or eliminating dependence on carbon-based energy, to redesign, recreate, and relocalize production while maintaining high quality of life. Transition projects therefore include reorganizing production and distribution of food, energy, transport, and housing, involving citizens in the design, implementation, and living of new practices.

Barry (2012: 27) argues that the transition town movement exemplifies a "creative adaptive management" route to more resilient and sustainable communities, in the "basic belief that communities of people can shape the conditions (socio-ecological and social) for their own flourishing" (p. 115) in "an integrated and seamless transformation of family, community and economic life" (Barry and Quilly, 2009: 15–16).

The transition towns movement is small, especially in comparison to the size of the problems it addresses; and particular transition towns vary enormously in their level of activity. But its focus on changing material relationships in the supply of basic needs, and on the inclusion of citizens in the reconstruction of a more sustainable everyday life, is shared by other movements. Rather than buying veggies flown across the globe (or even from the natural foods megamart), people get involved in community supported agriculture, collective gardening, urban farms, and sustainable cafés—transforming the relationship with the production, transportation, and consumption of food. Rather than simply cutting their own energy use, communities organize around the local development and networking of solar and wind energy, commuting plans that include more mass transit, biking and walking. And rather than just protesting working conditions in sweatshops, the disposability of fashion, alienation from technology, and unsustainable production, groups can turn to crafting, making, and mending, transforming means of production that are both alienating and unsustainable.

Countries where the supermarket, with foods shipped in from around the world, has been the longstanding indicator of material success (and excess), have seen growth in local farmers markets. In the US, the Department of Agriculture noted annual increases of around 16 percent in the five years to 2012, by when there were nearly 8000 (USDA, 2013). Millions of people engage in the growing, distribution, purchase, and consumption of locally grown—and often more sustainable—foods. Food markets, farms and gardens can also be found in urban areas. Detroit, for example, is a hotbed of urban agriculture and food justice; the Detroit Agriculture Network counts nearly 9000 urban gardens, has trained over 15,000 urban gardeners, and supports gardens, farms, and markets (Detroit Agriculture Network, 2013). Further developments include community-supported agriculture, where people buy shares in local farms; local (and less processed) foods in school lunches, including "farm to school" programs; local "food policy councils" (North American Food Policy Council, no date) to coordinate development of sustainable and secure practices; and ways for young people and students to learn to grow and prepare food sustainably. These moves challenge environmentally harmful industrialized agriculture.

We see similar, if to date less extensive, developments around the local production of energy. Examples include neighborhood solar installations, urban co- or tri-generation in green cities initiatives, and "just transitions" from coal and nuclear to wind and solar in the US southwest. Such developments illustrate not only the desire for post-carbon energy, but also an interest in being part of this transition in everyday life. Many cities around the globe have aggressive energy conservation programs as well as efforts to reduce sprawl and encourage more efficient modes of transport. To some extent, these local efforts are globally networked (ICLEI, no date).

These new materialist practices are ways for people to respond to their own sense of powerlessness by reclaiming the ability to shape their own lives. The food justice movement often refers to food *sovereignty* as a key goal (Alkon and Agyeman, 2011: 8). New materialism involves not only rejecting daily immersion in unsustainable and unjust relations and practices, but also creating alternatives that are embedded in bodies as they feed themselves and use energy. In becoming part of more sustainable flows of food, energy, and material, these movements express and embody empowerment.

Reconstructed flows of material needs can also be designed to pay attention to the processes and limitations of nature. As

Latour (2011) puts it, a reengaged movement would see "the process of human development as neither liberation from Nature nor as fall from it, but rather as a process of becoming ever-more attached to, and intimate with, a panoply of nonhuman creatures" (p. 17). For Barry (2012: 28), the transition movement tries to "render explicit those forms of relations of dependence on nature and fellow humans which have been occluded, forgotten, or otherwise hidden away in modernity. These include relations (material as well as symbolic) around food, the land, and the links between production, reproduction, and consumption." These movements are actively trying to replace separation with immersion in local environments, the domination of nature with recognition of the embedded material relationships we have with the nonhuman realm.

Critics in the green blogosphere (Read, 2008) disapprove of the transition town movement for its seemingly apolitical nature—it avoids electoral politics, and does not necessarily support those fighting polluters. However, efforts to advance sustainable and non-resilient material practices supplement a distant political effort that may or may not bear fruit with a local effort that puts one's body in the midst of actual change. Critics of the sustainable food movement point to the supposed elitism of some advocates (for example, Pollan, 2006)—they have income and knowledge to be able shift away from dominant practices. But there can also be a strong justice component in food movements, at least in groups that oppose racism, privilege, and the lack of quality food, while seeking access and power for the disadvantaged (Gottlieb and Joshi, 2010, Alkon and Agyeman, 2011).

What we can see in such movements is not then a post-material liberal interest group. Instead, they serve as examples of prefigurative politics as well as economics. Transition towns pre-figure or anticipate what is seen as an inevitable reversal of economic globalization (Barry and Quilly, 2009: 15). The point is not to lobby or vote for change, but rather literally to embody that change, and to illustrate alternative, more resilient, and more sustainable practices and relationships. As Maniates (2012: 121) puts it, "[i]n the search for a potent politics of transformation, everyday life may not be so everyday after all."

NEW FORMS OF GOVERNANCE

The new materialist movements we have surveyed are political as well as economic, but they do not actually say much about the governance aspects of politics. A new economy is going to need new forms of governance, ones not tied to the conventional economic growth imperative. We turn now to promising developments in polycentric, networked, and deliberative governance.

In 2009 Elinor Ostrom was awarded the Nobel Prize in Economics for her work on cooperative approaches to the local management of common resources. Ostrom (2009) and others have advocated a "polycentric" approach to climate governance that would involve the self-organization of mitigation efforts at multiple scales (downplaying global scale). The idea is that communities are quite capable of devising responses to collective action problems—such as climate change—without top-down management. While Ostrom's core evidence comes from work on localized common pool resources (fisheries, forests, and irrigation systems), she points to a few examples where (for example) cities have developed emission reduction plans, and sub-national units such as California have acted on climate change when national governments could not. The "energetic society" portrayed by Hajer (2011) (with specific reference to the Netherlands) shows that polycentrism can involve multiple initiatives by citizens, communities, cooperatives, and businesses, though Hajer also stresses the importance of a national government that facilitates experimentation, networking, and learning.

Polycentrism is consistent with our position in chapter 1 that responding to climate change requires many linked things, not one big thing. However, on Ostrom's own account, the conditions for effective polycentric governance prove demanding. The conditions include shared perception the common resource (climate system) is important, reliable information on the condition of the resource, trust among key actors, social capital (i.e. dense social networks involving public-spirited behavior), good communication about the state of the resource, effective monitoring of one another's behavior by those who access the resource, long time horizons, and linkage to individual benefits (such as lower fuel bills if renewable energy proves successful). None of these can be taken for granted. In treating climate change as mainly a problem of collective action (i.e. coordination of actors toward a common end), Ostrom's polycentric approach also ignores the divisive politics concerning especially who should bear the burden of action, the existence of powerful interests such as fossil fuel corporations that want to block action, and the growth-maximizing imperatives of local and national governments. None of this means a polycentric approach should be abandoned, and indeed we can point to local successes, not just in mitigation as stressed by Ostrom, but more straightforwardly in adaptation (for example, locally in Australia, where nationally mitigation policy has proven so divisive). We just need to figure out how the conditions Ostrom requires can actually be secured. Much will depend on changes in the terms of discourse and associated cultural dispositions. Polycentrism requires coordination across localized efforts; shared discourses featuring sustainability and climate justice rather than growth at all costs are one way of achieving this without central authority.

Networks are another way (see Chapter 6). Existing networks that organize carbon trading and technology transfer are populated mainly by actors conditioned by their history in established institutions, be they governments, international organizations, or markets. And often networks are intertwined with the old dominant forms. So the most prominent global networks have arisen for the trading of emissions, offsets, and technologies across national boundaries, and as such exist in tight association with markets. Within countries, governance networks are often overseen by national governments, and so do not exactly provide an alternative to these governments. What this means is that to date the performance of networks is not strikingly better than established forms of governance. But it does not have to stay that way, as new interests and actors emerge.

Hoffman (2011) treats emerging forms of climate governance as an "experimental system" that could expand in coherence, scope, and impact. Hoffman's new forms share three features. The first is "a common liberal environmental ethos that stresses the compatibility of economic growth and environmental protection" (p. 25)—what we would term ecological modernization. The second is a "market orientation" (p. 39). The third is their voluntary nature: nobody is compelled to join them so obstructive actors can simply be ignored. With time, Hoffman hopes that these experiments will create interests (for example, carbon traders) that can benefit from expansion, and also governance capacity (for example, in the certification of emissions trading and offsets). Though Hoffman does not put it in these terms, we suspect his hopes rest on actors socialized into new forms transcending their prior socialization in old forms of markets and government.

So can new forms escape the dead hand of the old? Of the 58 experiments Hoffman covers, only marginal ones such as transition towns (which do not in fact feature ecological modernization and a market orientation) seem likely to do so, and they look more like social movements than governance structures. But that very feature may actually provide a clue to how the performance of networked governance might be improved. As Braithwaite (2007: 167) puts it, networks need nodes of contestation, not just nodes of agreement. Radical social movements have parts to play in governance. Existing climate governance networks are dominated by moderate discourses, subscription to which may be the price of admission for activists and NGOs. However, without more profound contestation, networks will not necessarily produce results much different from established institutions. The same might be said about Ostrom's polycentric approach; effective governance is not just a matter of securing cooperation—it also requires moments of contestation in governance systems.

Of course the direction that contestation will induce depends a great deal on the content of the discourses that contestation introduces. When it comes to climate change, contestation from the direction of discourses emphasizing planetary boundaries, climate justice, and de-carbonization would be salutary. Scientists and other experts, individuals within government, or even ordinary citizens as well as activists could (and do) advance them.

Contestation could help turn networks into more productive *deliberative* systems by providing grist for deliberation. Deliberation here means that instead of being dominated by material self-interest or standard core priorities (economic, security, revenue, and welfare), politics would feature dialogue oriented to the production of mutually beneficial outcomes, persuasion rather than coercion, the pursuit of collective interests (such as global sustainability), reflection on what is desirable and defensible, and efforts to reach and understand those with different starting points. Deliberation in this sense could be sought in any kind of governance, be it global negotiations, minilateral forums of the kind we addressed in Chapter 6, local governments, partnerships, as well as networks. Deliberation can involve officials, activists, and ordinary citizens. Environmental affairs in general have featured many deliberative innovations in recent years (Smith, 2003). While deliberation is organized most straightforwardly at local levels, one of the world's biggest multinational efforts occurred in the lead-up to COP-15 in Copenhagen in 2009, when on September 26 100 ordinary citizens gathered in each of 38 countries to deliberate climate policy (Rask et al., 2012). In almost every country, the citizens supported both national and global action stronger than their governments were willing to adopt. The results of this "Word

Wide Views" process were presented in Copenhagen but had little impact on the official proceedings.

Making deliberative forums influential is a struggle. The IPCC (see Chapter 2) has deliberative aspects, but it is just one source of inputs into governance. The more difficult challenge involves reenvisioning whole political systems along deliberative lines. It is not exactly that key actors oppose the idea of deliberation, but rather that they are constrained by political roles they play— elected leader, negotiator, lobbyist, or uncompromising activist.

There are however practical steps that could help. In networked governance, what is missing is often the contestatory aspect of deliberation that could move networks out of the shadow of established power. The problem is that existing members of networks are unlikely to see any need for more contestation that would in the short term probably make their own lives more difficult. In which case, the burden of change falls upon outside activists. At the moment activists are often talking to each other in deliberative enclaves (symbolized physically by the Klimaforum of mostly radical activists in Copenhagen in 2009, and the Cochabamba "World People's Conference on Climate Change and the Rights of Mother Earth" in 2010). Activists themselves might resist closer engagement with existing networks on the grounds that this would compromise an agenda of radical transformation. NGOS such as the Climate Action Network sometimes try to bridge the gap, but they target UNFCCC negotiations rather than networks. The multilateral negotiations of UNFCCC could conceivably be restructured along deliberative lines. Currently those negotiations embody positional bargaining between states and blocks of states, under some conventional procedural rules. We know quite a lot about how to facilitate productive deliberation. Facilitators do this for a living; typically, they try to induce reciprocal understanding, a sense of shared problem-solving, willingness to contemplate common interests, listening to others.

What governance at all levels really needs is a "reflexive" capacity to contemplate its own reform. The experience of COP-15 at

Copenhagen in 2009, where the institutional deficiencies of the UNFCCC process led to years of painstaking effort by negotiators being overturned by an agreement patched together in a few hours by the US and China could have led negotiators to recognize the need for a global reflexive capacity; but that is not what happened. At the margins we do see reflexivity among those searching for new forms of governance as alternatives to the failures of states and the UNFCCC process. The puzzle remains concerning how a more general global capacity might develop when not spurred by a compelling recent catastrophe such as total war.

We still lack obvious positive examples of forms of climate governance that are good enough on a scale big enough. Yet we know (see Chapter 6) that some governance arrangements do better than others. We know plenty about the reasons for failure, and in theory quite a bit about (deliberative and other) arrangements that ought to work better. We have plenty of examples of new forms of governance, especially those involving networks and public–private partnerships. We have some local success stories. In short, we have plenty to build upon when it comes to reconstructing governance for a climate-challenged society.

CONCLUSIONS

In Chapter 1 we argued that a climate-challenged society is going to require many linked initiatives, rather than one big response. We have explained what these multiple initiatives might involve, though we still need creative thinking about how to combine the impetus for radical change with what seems feasible in the context of the dominant political economy. That political economy is not going to give way easily, no matter how compelling any case grounded in the realities of climate change. There remain some very big questions. Should we try to transform existing growth-at-any-costs capitalism into effective climate capitalism? Or should we be thinking about new kinds of economies that transcend the growth imperative? Is it worth trying to redeem existing dominant forms of national government and the global climate regime by asking them to perform better, or should we abandon them in favor of new pluralistic and networked forms? There are no easy answers to such big questions, but there are better and worse ways of interrogating them. In our canvassing of discursive ways to connect scientific knowledge to a democratic society, in thinking about the limits to conventional economics and what that tells us about confronting ethical issues head-on, in looking at the variety of available actions and instruments for mitigation and adaptation, in showing the necessity of tackling climate justice, in exposing the deficiencies of existing governance forms and examining the promise of new kinds of governance, and in demonstrating that we now need to think in co-evolutionary and ecologically rational terms about how to negotiate the Anthropocene, we hope to have provided plenty of resources for intelligent thinking. In an ever-evolving climate-challenged yet necessarily pluralistic society, nobody gets to have the last word—and that applies to ourselves as much as anyone else.

Bibliography

Adger, W. Neil. 2003. Social Capital, Collective Action and Adaptation to Climate Change. *Economic Geography* 79 (4): 387–404.

Adger, W. Neil, Suraje Dessai, Marisa Goulden, Mike Hulme, Irene Lorenzoni, Donald R. Nelson, Lars Otto Naess, Johanna Wolf, and Anita Wreford. 2009. Are There Social Limits to Adaptation to Climate Change? *Climatic Change* 93: 335–54.

Adger, W. Neil, Katrina Brown, Donald R. Nelson, Fikret Berkes, Hallie Eakin, Carl Folke, Kathleen Galvin, Lance Gunderson, Marisa Goulden, Karen O'Brien, Jack Ruitenbeek, and Emma L. Tompkins. 2011a. Resilience Implications of Policy Responses to Climate Change. *WIREs Climate Change* 2: 757–66.

Adger, W. Neil, Katrina Brown, and James Waters. 2011b. Resilience. In John S. Dryzek, Richard B. Norgaard, and David Schlosberg, eds, *The Oxford Handbook of Climate Change and Society*. Oxford: Oxford University Press, pp. 696–710.

Agarwal, Anil, Sunita Narain, and Anju Sharma. 2002. The Global Commons and Environmental Justice—Climate Change. In John Byrne, Leigh Glover, and Celia Martinez, eds, *Environmental Justice: International Discourses in Political Economy—Energy and Environmental Policy*. Piscatawny, NJ: Transaction Publishers, pp. 171–201.

Agrawala, Shardul. 2011. PLAN or REACT? Analysis of Adaptation Costs and Benefits Using Integrated Assessment Models. *Climate Change Economics* 2 (3): 1–36.

Alkon, Alison and Julian Agyeman. 2011. *Cultivating Food Justice*. Cambridge, MA: MIT Press.

Bäckstrand, Karin. 2010. The Legitimacy of Global Public-Private Partnerships on Climate and Sustainable Development. In Karin Bäckstrand, Jamil Khan, Annica Kronsell, and Eva Lövbrand, eds, *Environmental Politics and Deliberative Democracy: Examining the Promise of New Modes of Governance*. Cheltenham: Edward Elgar, pp. 85–104.

Baer, Paul. 2011. International Justice. In John S. Dryzek, Richard B. Norgaard, and David Schlosberg, eds, *The Oxford Handbook of Climate Change and Society*. Oxford: Oxford University Press, pp. 323–35.

Barnett, Jon. 2011. Human Security. In John S. Dryzek, Richard B. Norgaard, and David Schlosberg, eds, *The Oxford Handbook of Climate Change and Society*. Oxford: Oxford University Press, pp. 267–77.

Barnosky, Anthony D., Elizabeth A. Hadly, Jordi Bascompte, Eric L. Berlow, James H. Brown, Mikael Fortelius, Wayne M. Getz, John Harte, Alan Hastings, Pablo A. Marquet, Neo D. Martinez, Arne Mooers, Peter Roopnarine, Geerat Vermeij, John W. Williams, Rosemary Gillespie, Justin Kitzes, Charles Marshall, Nicholas Matzke, David P. Mindell, Eloy Revilla, and Adam B. Smith. 2012. Approaching a State Shift in Earth's Biosphere. *Nature* 486: 52–58.

Barry, John. 2012. *The Politics of Actually Existing Unsustainability*. Oxford: Oxford University Press.

Barry, John and Steve Quilley. 2009. The Transition to Sustainability: Transition Towns and Sustainable Communities. In Liam Leonard and John Barry, eds, *The Transition to Sustainable Living and Practice*. Cheltenham: Emerald, pp. 1–28.

Beeson, Mark. 2010. The Coming of Environmental Authoritarianism. *Environmental Politics* 19 (2): 276–94.

Bell, Derek. 2011. Does Anthropogenic Climate Change Violate Human Rights? *Critical Review of International Social and Political Philosophy* 14 (2): 99–124.

Berry, Helen L., Anthony Hogan, Jennifer Owen, Debra Rickwood, and Lyn Fragar. 2011. Climate Change and Farmers' Mental Health: Risks and Responses. *Asia-Pacific Journal of Public Health* 23 (2 suppl): 119S–132S.

Benedick, Richard E. 1991. *Ozone Diplomacy: New Directions in Saving the Planet*. Cambridge, MA: Harvard University Press.

Benyus, Janine. 1997. *Biomimicry: Innovation Inspired by Nature*. New York: William Morrow.

Berejikian, Jeffrey. 2004. *International Relations Under Risk: Framing State Choice*. Albany: State University of New York Press.

Bhattacharya, S. C. and Jana Chimnoy. 2009. Renewable Energy in India: Historical Development and Prospects. *Energy 34* (8): 981–91.

Biermann, Frank. 2011. New Actors and Mechanisms of Global Governance. In John S. Dryzek, Richard B. Norgaard, and David Schlosberg, eds, *The Oxford Handbook of Climate Change and Society*. Oxford: Oxford University Press, pp. 685–95.

Biermann, Frank, K. Abbott, S. Andresen, K. Bäckstrand, S. Bernstein, M. M. Betsill, H. Bulkeley, B. Cashore, J. Clapp, C. Folke, A. Gupta,

J. Gupta, P. M. Haas, A. Jordan, N. Kanie, T. Kluvánková-Oravská, L. Lebel, D. Liverman, J. Meadowcroft, R. B. Mitchell, P. Newell, S. Oberthür, L. Olsson, P. Pattberg, R. Sánchez-Rodríguez, H. Schroeder, A. Underdal, S. Camargo Vieira, C. Vogel, O. R. Young, A. Brock, and R. Zondervan. 2012. Navigating the Anthropocene: Improving Earth System Governance. *Science* 335: 1306–7.

Blashki, Grant, Greg Armstrong, Helen Louise Berry, Haylee J. Weaver, Elizabeth G. Hanna, Peng Bi, David Harley, and Jeffery Thomas Spickett. 2011. Preparing Health Services for Climate Change in Australia. *Asia-Pacific Journal of Public Health* 23 (2 suppl): 133S–143S.

Bok, Derek. 2010. *The Politics of Happiness: What Government Can Learn from the New Research on Well-Being*. Princeton: Princeton University Press.

Bosello, Francesco, Carlo Carraro, and Enrica De Cian. 2012. Market and Policy Driven Adaptation. Paper prepared for the Copenhagen Consensus 2012. Available at <http://copenhagenconsensus.com/sites/default/files/Climate%2BAdaptation.pdf>.

Boykoff, Maxwell T. and Jules M. Boykoff. 2004. Balance as Bias: Global Warming and the US Prestige Press. *Global Environmental Change* 14: 125–36.

Braithwaite, Chris. 2010. *Scenario Planning and Sensitivity Analysis*. Melbourne: KPMG.

Braithwaite, John. 2007. Contestatory Citizenship, Deliberative Denizenship. In Geoffrey Brennan, Frank Jackson, Robert Goodin, and Michael Smith, eds, *Common Minds*. Oxford: Oxford University Press, pp. 161–81.

Broome, John. 2012. *Climate Matters: Ethics in a Warming World*. New York: W. W. Norton.

Bulkeley, Harriett and Michelle Betsill. 2003. *Cities and Climate Change: Urban Sustainability and Global Environmental Governance*. London: Routledge.

Burton, Ian. 1994. Deconstructing Adaptation (. . .) and Reconstructing. *Delta* 5 (1): 14–15.

Caney, Simon. 2005. Cosmopolitan Justice, Responsibility, and Global Climate Change. *Leiden Journal of International Law* 18 (4): 747–75.

Caney, Simon. 2010. Climate Change, Human Rights, and Moral Thresholds. In Stephen Gardiner, Simon Caney, Dale Jamieson, and Henry Shue, eds, *Climate Ethics*. Oxford: Oxford University Press.

Cannon, Terry and Detlef Müller-Mahn. 2010. Vulnerability, Resilience and Development Discourses in Context of Climate Change. *Natural Hazards* 55: 621–35.

Catney, Philip and Timothy Doyle. 2011. The Welfare of Now and the Green (Post) Politics of the Future. *Critical Social Policy* 31 (2): 174–93.

Chakravarty, Shoibal, Ananth Chikkatur, Heleen de Coninck, Stephen Pacala, Robert Socolow, Massimo Tavoni. 2009. Sharing Global CO2 Emission Reductions among One Billion High Emitters. *Proceedings of the National Academy of Sciences* 106: 11885–8.

Chakravarty, Sukhamoy. 1987. Cost–Benefit Analysis. In John Eatwell, Murray Milgate, and Peter Newman, eds, *The New Palgrave: A Dictionary of Economics,* vol. I. London: Macmillan, pp. 687–90.

Chester, Charles C., Jodi A. Hilty, and Stephen C. Trombulak. 2012. Climate Change Science, Impacts, and Opportunities. In Jodi A. Hilty, Charles C. Chester, and Molly S. Cross eds, *Climate and Conservation: Landscape and Seascape Science, Planning, and Action.* Washington, DC: Island Press, pp. 3–16.

Christoff, Peter and Robyn Eckersley. 2011. Comparing State Responses. In John S. Dryzek, Richard B. Norgaard, and David Schlosberg, eds, *The Oxford Handbook of Climate Change and Society.* Oxford: Oxford University Press, pp. 431–48.

Cinner, Joshua E., T. R. McClanahan, N. A. J. Graham, T. M. Daw, J. Maina, S. M. Stead, A. Wamukota, K. Brown, and Ö. Bodin. 2012. Vulnerability of Coastal Communities to Key Impacts of Climate Change on Coral Reef Fisheries. *Global Environmental Change* 22 (1): 12–20.

Climate Justice Now! 2011. Climate Justice Groups Call for Binding Deep and Drastic GHG Emissions Cuts by Developed Countries. Available at <http://www.climate-justice-now.org/climate-justice-groups-call-for-binding-deep-and-drastic-ghg-emissions-cuts-by-developed-countries/>.

Cournot, Agustin. 1838. *Recherches sur les Principes Mathématiques de la Théorie des Richesses.* Paris: Chez L. Hachette.

Cronon, William. 1995. The Trouble with Wilderness, or Getting Back to the Wrong Nature. In William Cronon, ed., *Uncommon Ground: Rethinking the Human Place in Nature.* New York: W. W. Norton, pp. 69–90.

Crutzen, Paul. 2002. Geology of Mankind. *Nature* 415 (3): 23.

Crutzen, Paul. 2006. Albedo Enhancement by Stratospheric Sulphur Injections: A Contribution to Resolve a Policy Dilemma? *Climatic Change* 77: 211–19.

Crutzen, Paul. 2010. The New World of the Anthropocene. *Environmental Science and Technology* 44 (7): 2228–31.

Crutzen, Paul, and E. F. Stoermer. 2000. The Anthropocene. *Global Change Newsletter* 41: 17–18.

DeCanio, Stephen J. 2003. Economic Analysis, Environmental Policy, and Intergenerational Justice in the Reagan Administration: The Case of the Montreal Protocol. *International Environmental Agreements: Politics, Law and Economics* 3(4): 299–321.

Detroit Agriculture Network. 2013. Keep Growing Detroit. Available at <http://detroitagriculture.net/>.

Diamond, Jared. 2005. *Collapse: How Societies Choose to Fail or Survive.* New York: Viking Penguin.

Dryzek, John S. 1987. *Rational Ecology: Environment and Political Economy.* New York: Basil Blackwell.

Dryzek, John S., David Downes, Christian Hunold, David Schlosberg with Hans-Kristian Hernes. 2003. *Green States and Social Movements: Environmentalism in the United States, United Kingdom, Germany, and Norway.* Oxford: Oxford University Press.

Easterlin, Richard A. 1973. Does Money Buy Happiness? *The Public Interest* 30 (10): 3–10.

Easterlin, Richard A., Laura Angelescu McVey, Malgorzata Switek, Onnicha Sawangfa, and Jacqueline Smith Zweig. 2010. The Happiness-Income Paradox Revisited. *Proceedings of the National Academy of Sciences* 107 (52): 22463–8.

Eckersley, Robyn. 2012. Moving Forward in the Climate Negotiations: Multilateralism or Minilateralism? *Global Environmental Politics* 12 (2): 24–42.

EcoEquity. 2008. Greenhouse Development Rights. Available at <www.gdrights.org>.

Economist. 2011a. A Man-Made World: Science is Recognizing Humans as a Geological Force to be Reckoned With. May 26.

Economist. 2011b. Welcome to the Anthropocene. May 26.

Edwards, Paul N. 2010. *A Vast Machine: Computer Models, Climate Data, and the Politics of Global Warming.* Cambridge, MA: MIT Press.

EJCCI (Environmental Justice and Climate Change Initiative). 2002. *10 Principles for Just Climate Change Policies in the U.S.* Available at <http://www.ejcc.org/ejcc10short_usa.pdf>.

Fisher-Vanden, Karen, Ian Sue Wing, Elisa Lanzi and David C. Popp. 2011. Modeling Climate Change Adaptation: Challenges, Recent Developments and Future Directions. Available at <http://people.bu.edu/isw/>.

Folke, Carl. 2006. Resilience: The Emergence of a Perspective for Social-Ecological Systems Analysis. *Global Environmental Change* 16: 253–67.

Folke, Carl, Åsa Jansson, Johan Rockström, Per Olsson, Stephen R. Carpenter, F. Stuart Chapin III, Anne-Sophie Crépin, Gretchen

Daily, Kjell Danell, Jonas Ebbesson, Thomas Elmqvist, Victor Galaz, Fredrik Moberg, Måns Nilsson, Henrik Österblom, Elinor Ostrom, Åsa Persson, Garry Peterson, Stephen Polasky, Will Steffen, Brian Walker, and Frances Westley. 2011. Reconnecting to the Biosphere. *Ambio* 40: 719–738.

Gardiner, Stephen M. 2011. *A Perfect Moral Storm: The Ethical Tragedy of Climate Change.* Oxford: Oxford University Press.

Garnaut, Ross. 2008. *The Garnaut Climate Change Review.* Cambridge: Cambridge University Press.

Gilman, Nils, Doug Randall, and Peter Schwartz. 2011. Climate Change and "Security." In John S. Dryzek, Richard B. Norgaard, and David Schlosberg, eds, *The Oxford Handbook of Climate Change and Society.* Oxford: Oxford University Press, pp. 251–66.

Glacken, Clarence J. 1967. *Traces on the Rhodian Shore.* Berkeley, CA: University of California Press.

Gottlieb, Robert and Anupama Joshi. 2010. *Food Justice.* Cambridge, MA: MIT Press.

Gottlieb, Roger S. 2006. *A Greener Faith: Religious Environmentalism and Our Planet's Future.* Oxford: Oxford University Press.

Guardian. 2010. Interview with James Lovelock. Available at <http://www.guardian.co.uk/environment/2010/mar/29/james-lovelock>.

Guardian. 2012. Interview with Peter Bakker. Available at, <http://www.guardian.co.uk/sustainable-business/rio-20-business-sustainable-development>. June 22.

Hajer, Maarten. 2011. *The Energetic Society: In Search of a Governance Philosophy for a Clean Economy.* The Hague: PBL Netherlands Environmental Assessment Agency.

Hall, Peter A. and David Soskice, eds. 2001. *Varieties of Capitalism.* Oxford: Oxford University Press.

Hanemann. W. Michael. 2008. *What is the Economic Cost of Climate Change?* Berkeley, CA: University of California Department of Agricultural and Resource Economics.

Hanna, Elizabeth G. 2011. Health Hazards. In John S. Dryzek, Richard B. Norgaard, and David Schlosberg, eds, *The Oxford Handbook of Climate Change and Society.* Oxford: Oxford University Press, pp. 217–31.

Hansen, James. 2009. *Storms of my Grandchildren: The Truth About the Coming Climate Catastrophe and our Last Chance to Save Humanity.* New York: Bloomsbury.

Hare, Bill and Malte Meinshausen. 2006. How Much Warming are we Committed to and How Much Can be Avoided? *Climatic Change* 75: 111–49.

Higgs, Eric. 2003. *Nature by Design.* Cambridge, MA: MIT Press.

Higgs, Eric. 2012. History, Novelty, and Virtue in Ecological Restoration. In Allen Thompson and Jeremy Bendik-Keymer, eds, *Ethical Adaptation to Climate Change: Human Virtues of the Future*. Cambridge, MA: MIT Press, pp. 81–102

Hobson, Kersty and Simon Niemeyer. 2013. What Sceptics Believe: The Effects of Information and Deliberation on Climate Scepticism. *Public Understanding of Science* 22: 396–412.

Hoffman, Matthew J. 2011. *Climate Governance at the Crossroads: Experimenting with a Global Response after Kyoto*. Oxford: Oxford University Press.

Holland, Breena. 2008. Justice and the Environment in Nussbaum's "Capabilities Approach": Why Sustainable Ecological Capacity Is a Meta-Capability. *Political Research Quarterly* 61 (2): 319–32.

Holland, Breena. 2012. Environment as Meta-Capability: Why a Dignified Human Life Requires a Stable Climate System. In Allen Thompson and Jeremy Bendik-Keymer, eds, *Ethical Adaptation to Climate Change: Human Virtues of the Future*. Cambridge, MA: MIT Press, pp. 145–64.

Holm, Poul, Anne Husum Marboe, Bo Poulsen, and Brian R. MacKenzie. 2010. Marine Animal Populations: A New Look Back In Time. In Alasdair McIntyre, ed., *Life in the World's Oceans: Diversity, Distribution, and Abundance*. Oxford: Basil Blackwell, pp. 3–23.

Howarth, Richard B. and Norgaard, Richard B. 1992. Environmental Valuation Under Sustainable Development. *American Economic Review* 82 (2): 473–7.

Hulme, Mike. 2009. *Why We Disagree About Climate Change: Understanding Controversy, Inaction and Opportunity*. Cambridge: Cambridge University Press.

Hunt, Alistair and Paul Watkiss. 2011. Climate Change Impacts and Adaptation in Cities: a Review of the Literature. *Climatic Change* 104 (1): 13–49.

ICLEI. No date. Home. ICLEI, Available at <http://www.iclei.org>.

Ikeme, Jekwu. 2003. Equity, Environmental Justice And Sustainability: Incomplete Approaches In Climate Change Politics. *Global Environmental Change* 13: 195–206.

Inglehart, Ronald. 1997. *Modernization and Postmodernization*. Princeton, NJ: Princeton University Press.

International Climate Justice Network (2002). Bali Principles of Climate Justice. Available at <http://www.corpwatch.org/article.php?id=3748>.

Jackson, Tim. 2009. *Prosperity without Growth: Economics for a Finite Planet*. London: Earthscan.

Jamieson, Dale. 2011. The Nature of the Problem. In John S. Dryzek, Richard B. Norgaard, and David Schlosberg, eds, *The Oxford Handbook of Climate Change and Society*. Oxford, UK: Oxford University Press, pp. 38–54.

Jasanoff, Sheila. 2011. Cosmopolitan Knowledge: Climate Science and Global Civic Epistemology. In John S. Dryzek, Richard B. Norgaard, and David Schlosberg, eds, *The Oxford Handbook of Climate Change and Society*. Oxford: Oxford University Press, pp. 129–43.

Jasanoff, Sheila. 2010. A New Climate for Society. *Theory, Culture and Society* 27: 233–53.

Jordan, Andrew, Dave Huitema, Harro van Asselt, Tim Rayner, and Frans Berkhout, eds. 2010. *Climate Change Policy in the European Union: Confronting the Dilemmas of Mitigation and Adaptation?* Cambridge: Cambridge University Press.

Kahan, Dan M., Hank Jenkins-Smith, and Donald Braman. 2010. Cultural Cognition of Scientific Consensus. *Journal of Risk Research* 14 (2): 147–74.

Kearns, Laurel. 2011. The Role of Religions in Activism. In John S. Dryzek, Richard B. Norgaard, and David Schlosberg, eds, *The Oxford Handbook of Climate Change and Society*. Oxford: Oxford University Press, pp. 414–28.

Keith, David. 2000. Geoengineering: History and Prospect. *Annual Review of Energy and the Environment 25:* 245–84.

Kelly, P. Mick and W. Neil Adger. 2000. Theory and Practice in Assessing Vulnerability to Climate Change and Facilitating Adaptation. *Climatic Change* 47 (4): 325–52.

Kern, Florian and Michael Howlett. 2009. Implementing Transition Management as Policy Reforms: A Case Study of the Dutch Energy Sector. *Policy Sciences* 42: 391–408.

Klinenberg, Eric. 2003. *Heat Wave: A Social Autopsy of Disaster in Chicago*. Chicago: University of Chicago Press.

Kurzweil, Ray. 2005. *The Singularity is Near*. London: Penguin.

Latour, Bruno. 2011. Love Your Monsters. In Michael Shellenberger and Ted Nordhaus, eds, *Love Your Monsters: Postenvironmentalism and the Anthropocene*. Oakland, CA: Breakthrough Institute, pp. 16–23

Liao, S. Matthew, Anders Sandberg, and Rebecca Roache. 2012. Human Engineering and Climate Change. *Ethics, Policy, and the Environment* 15 (2): 206–21.

Light, Andrew. 2011. Six Reasons Why the Durban Decision Matters. *Think Progress blog*. Available at <http://thinkprogress.org/climate/2011/12/18/391533/six-reasons-why-the-durban-decision-matters/>.

Litfin, Karen T. 1994. *Ozone Discourses: Science and Politics in Global Environmental Cooperation.* New York: Columbia University Press.

Lo, Alex. 2011. Deliberative Monetary Valuation as a Political-Economic Methodology: Exploring the Prospects for Value Pluralism with a Case Study on Australian Climate Change Policy. PhD thesis, Australian National University.

Lomborg, Bjørn. 2007. *Cool It! The Skeptical Environmentalist's Guide to Global Warming.* New York: Knopf.

Lomborg, Bjørn. 2010. Introduction. In Bjørn Lomborg, ed., *Smart Solutions to Climate Change: Comparing Costs and Benefits.* Cambridge: Cambridge University Press, pp. 1–5.

Lovejoy, Thomas. (2012), The Greatest Challenge of Our Species. *New York Times*, April 5.

Maniates, Michael F. 2012. Everyday Possibilities. *Global Environmental Politics* 12 (1): 121–5.

Mann, Michael. 2012. *The Hockey Stick and the Climate Wars: Dispatches from the Front Lines.* New York: Columbia University Press.

Mann, Michael E., Raymond S. Bradley, and Malcolm K. Hughes. 1998. Global-Scale Temperature Patterns and Climate Forcing over the Past Six Centuries. *Nature* 392: 779–87.

McCarthy, James J, Osvaldo F. Canziani, Neil A. Leary, David J. Dokken, and Kasey S. White, eds. 2001. *Climate Change 2001: Impacts, Adaptation and Vulnerability. Contribution of Working Group II to the Third Assessment Report of the Intergovernmental Panel on Climate Change.* Cambridge: Cambridge University Press.

McCright, Aaron and Riley Dunlap. 2011. The Politicization of Climate Change and Polarization in the American Public's Views of Global Warming, 2001–2010. *Sociological Quarterly* 52: 155–94.

McKibben, Bill. 2006 [1989]. *The End of Nature.* New York: Random House.

Meadowcroft, James and Oluf Langhelle. 2009. *Caching the Carbon: The Politics and Policy of Carbon Capture and Storage.* Cheltenham: Edward Elgar.

Meadows, Donella H., Dennis L. Meadows, Jørgen Randers, and William W. Behrens III. 1972. *The Limits to Growth: A Report for the Club of Rome's Project on the Predicament of Mankind.* New York. Universe Books.

Measham, Thomas G., Benjamin L. Preston, Timothy F. Smith, Cassandra Brooke, Russell Gorddard, Geoff Withycombe, and Craig Morrison. 2010. *Adapting to Climate Change Through Local Municipal Planning: Barriers and Opportunities.* Canberra: CSIRO Working Paper Series, 25.

Meinshausen, Malte Nicolai, William Hare, Sarah C. B. Raper, Katja Frieler, Reto Knutti, David J. Frame, and Myles R. Allen. 2009. Greenhouse-gas emission targets for limiting global warming to 2 C. *Nature* 458 (issue 7242): 1158–62.

Mendelsohn, Robert. 2008. Is the Stern Review an Economic Analysis? *Review of International Economics and Policy* 2 (1): 45–60.

Moser, Susanne C. and Lisa Dilling, eds. 2007. *Creating a Climate for Change: Communicating Climate Change and Facilitating Social Change.* Cambridge: Cambridge University Press.

Moser, Susanne C. and Lisa Dilling. 2011. Communicating Climate Change. In John S. Dryzek, Richard B. Norgaard, and David Schlosberg, eds, *The Oxford Handbook of Climate Change and Society.* Oxford: Oxford University Press, pp. 161–74.

Naím, Moisés. 2009. Minilateralism. *Foreign Policy* 173 (July/August): 135–6.

Nelson, Robert H. 1987. The Economics Profession and Public Policy. *Journal of Economic Literature* 25 (1): 49–91.

New York Times. 2012. Profits on Carbon Credits Drive Output of a Harmful Gas, *New York Times*, 9 August.

Newell, Peter and Matthew Paterson. 2010. *Climate Capitalism: Global Warming and the Transformation of the Global Economy.* Cambridge: Cambridge University Press.

Nordhaus, William D. 1990. Greenhouse Economics: Count Before You Leap. *The Economist* July 7, p. 21.

Nordhaus, William D. 1993. Rolling the "DICE": An Optimal Transition Path for Controlling Greenhouse Gases. *Resource and Energy Economics* 15: 27–50.

Nordhaus, William D. 2007a. A Review of the Stern Review on the Economics of Climate Change. *Journal of Economic Literature* 45 (3): 686–702.

Nordhaus, William D. 2007b. To Tax or Not to Tax: Alternative Approaches to Slowing Global Warming. *Review of Environmental Economics and Policy* 1 (1): 26–44.

Nordhaus, William. D. and Joseph Boyer. 2000. *Warming the World: Economic Models of Global Warming.* Cambridge, MA: MIT Press.

Norgaard, Kari Marie. 2011. *Living in Denial: Climate Change, Emotions, and Everyday Life.* Cambridge, MA: MIT Press.

Norgaard, Richard B. 1988. Sustainable Development: A Co-evolutionary View. *Futures* 20 (6): 606–20.

North American Food Policy Council. No date. What is a Food Policy Council? Available at <http://www.foodsecurity.org/FPC/index.html>.

Nussbaum, Martha C. 2006. *Frontiers of Justice: Disability, Nationality, Species Membership*. Cambridge, MA: Harvard University Press.

Oreskes, Naomi and Erik Conway. 2010. *Merchants of Doubt: How a Handful of Scientists Obscured the Truth from Tobacco to Global Warming*. New York: Bloomsbury Press.

Ostrom, Elinor. 2009. A Polycentric Approach for Coping with Climate Change. World Bank Policy Research Working Paper 5095. Washington DC: World Bank.

Pacala, Stephen and Robert Socolow. 2004. Stabilization Wedges: Solving the Climate Problem for the Next 50 Years with Current Technologies. *Science* 305, No. 5686 (13 August): 968–72.

Paterson, Matthew. 2011. Selling Carbon: From International Climate Regime to Global Carbon Market. In John S. Dryzek, Richard B. Norgaard, and David Schlosberg, eds, *The Oxford Handbook of Climate Change and Society*. Oxford: Oxford University Press, pp. 611–24.

Pattberg, Philipp. 2010. The Role and Relevance of Networked Climate Governance. In Frank Biermann, Phiipp Pattberg, and Fariborz Zelli, eds, *Climate Governance Beyond 2012: Architecture, Agency and Adaptation*. Cambridge: Cambridge University Press, pp. 146–64.

Pelling, Mark. 2011. *Adaptation to Climate Change: From Resilience to Transformation*. London: Routledge.

Phillips, Leigh. 2012. Sea Versus Senators: North Carolina Sea-Level Rise Accelerates while State Legislators Put the Brakes on Research. *Nature* online, available at <http://www.nature.com/news/sea-versus-senators-1.10893>.

Pogge, Thomas. 2002. *World Poverty and Human Rights*. Cambridge: Polity Press.

Pollan, Michael. 2006. *Omnivore's Dilemma*. New York: Penguin.

Posner, Eric and David Weinbach. 2010. *Climate Change Justice*. Princeton: Princeton University Press.

Putnam, Robert. 2000. *Bowling Alone: The Collapse and Revival of American Community*. New York: Simon and Schuster.

Rask, Mikko, Richard Worthington, and Minna Lammi, eds. 2012. *Citizen Participation in Global Environmental Governance*. London: Earthscan.

Read, Rupert. 2008. Transition Towns are Great—But They Won't Save Us, Without Help. Available at, <http://www.rupertsread.blogspot.com.au/2008/02/transition-towns-are-great-but-they.html>.

Reid, Julian. 2012. The Disastrous and Politically Debased Subject of Resilience. *Development Dialogue* 58: 67–80.

Resilience Alliance. 2013. Resilience. Available at, <http://www.resalliance.org/index.php/resilience>.

Rittel, Horst W.J. and Melvin M. Webber. 1973. Dilemmas in a General Theory of Planning. *Policy Sciences* 4 (2): 155–69.

Rockström, Johan, Will Steffen, Kevin Noone, Åsa Persson, F. Stuart Chapin, III, Eric F. Lambin, Timothy M. Lenton, Marten Scheffer, Carl Folke, Hans Joachim Schellnhuber, Björn Nykvist, Cynthia A. de Wit, Terry Hughes, Sander van der Leeuw, Henning Rodhe, Sverker Sörlin, Peter K. Snyder, Robert Costanza, Uno Svedin, Malin Falkenmark, Louise Karlberg, Robert W. Corell, Victoria J. Fabry, James Hansen, Brian Walker, Diana Liverman, Katherine Richardson, Paul Crutzen, and Jonathan A. Foley. 2009. A Safe Operating Space for Humanity. *Nature* 461 (24 Sept): 472–5.

Rogelj, Joeri and Malte Meinshausen. 2010. Copenhagen Accord Pledges are Paltry. *Nature* 464 (22 April): 1126–28.

Sagoff, Mark. 2011. The Poverty of Climate Economics. In John S. Dryzek, Richard B. Norgaard, and David Schlosberg, eds, *The Oxford Handbook of Climate Change and Society*. Oxford: Oxford University Press, pp. 55–66.

Sandler, Ronald. 2012. Global Warming and Virtues of Ecological Restoration. In Allen Thompson and Jeremy Bendik-Keymer, eds, *Ethical Adaptation to Climate Change: Human Virtues of the Future*. Cambridge, MA: MIT Press, pp. 63–80.

Sarewitz, Daniel. 2004. How Science Makes Environmental Controversies Worse. *Environmental Science and Policy* 7: 385–403.

Schelling, Thomas. 1992. Some Economics of Global Warming. *American Economic Review* 82 (1): 1–14.

Schlosberg, David. 2012. Climate Justice and Capabilities: A Framework for Adaptation Policy. *Ethics and International Affairs 26* (4): 445–61.

Schneider, Stephen. 2009. *Science as a Contact Sport: Inside the Battle to Save Earth's Climate*. Washington, DC: National Geographic Society.

Sen, Amartya. 1999. *Development as Freedom*. New York: Anchor.

Shue, Henry. 1992. The Unavoidability of Justice. In Andrew Hurrell and Benedict Kingsbury, eds, *The International Politics of the Environment*. Oxford: Oxford University Press.

Shue, Henry. 1993. Subsistence Emissions and Luxury Emissions. *Law and Policy* 15: 39–59.

Shue, Henry. 1999. Global Environment and International Inequality. *International Affairs* 75: 533–7.

Silver, Lee M. 1997. *Remaking Eden: How Genetic Engineering and Cloning will Transform the American Family*. New York: Harper Collins.

Singer, Peter. 2004. *One World: The Ethics of Globalization*. New Haven CT: Yale University Press.

Smith, Graham. 2003. *Deliberative Democracy and the Environment.* London: Routledge.

Society for Ecological Restoration. 2004. *International Primer on Ecological Restoration.* Available at <http://www.ser.org/resources/resources-detail-view/ser-international-primer-on-ecological-restoration>.

Spash, Clive L. 2010. The Brave New World of Carbon Trading. *New Political Economy* 15 (2): 169–95.

Speth, James Gustave. 2012. *America the Possible: A Manifesto for a New Economy.* New Haven: Yale University Press.

Spickett, Jeffrey T., Helen L. Brown, and Dianne Katscherian. 2011. Adaptation Strategies for Health Impacts of Climate Change in Western Australia: Application of a Health Impact Assessment Framework. *Environmental Impact Assessment Review* 31 (3): 297–300.

Srinivasan, U. Thara, Susan P. Carey, Eric Hallstein, Paul A. T. Higgins, Amber C. Kerr, Laura E. Koteen, Adam B. Smith, Reg Watson, John Harte, and Richard B. Norgaard. 2008. The Debt of Nations and the Distribution of Ecological Impacts from Human Sctivity. *Proceedings of the National Academy of Sciences* 105 (5): 1768–73.

Stavins, Robert N. 2009. Cap and Trade or a Carbon Tax? *Environmental Forum* 25 (1): 16.

Steffen, Will. 2011. A Truly Complex and Diabolical Policy Problem. In *The Oxford Handbook of Climate Change and Society*, John S. Dryzek, Richard B. Norgaard, and David Schlosberg, eds, Oxford: Oxford University Press, pp. 21–37.

Steffen, Will, Paul J. Crutzen, and John R. McNeill. 2007. The Anthropocene: Are Humans Now Overwhelming the Great Forces of Nature? *Ambio* 36 (8): 614–21.

Steffen, Will, Jacques Grinevald, Paul Crutzen, and John McNeil. 2011a. The Anthropocene: Conceptual and Historical Perspectives. *Philosophical Transactions of the Royal Society* 369: 842–67.

Steffen, Will, Åsa Persson, Lisa Deutsch, Jan Zalasiewicz, Mark Williams, Katherine Richardson, Carole Crumley, Paul Crutzen, Carl Folke, Line Gordon, Mario Molina, Veerabhadran Ramanathan, Johan Rockström, Marten Scheffer, Hans Joachim Schellnhuber, and Uno Svedin. 2011b. The Anthropocene: From Global Change to Planetary Stewardship. *AMBIO: A Journal of the Human Environment* 40 (7): 738–761.

Stern, Nicholas. 2007. *The Economics of Climate Change: The Stern Review.* Cambridge: Cambridge University Press.

Stern, Nicholas. 2013. I Got it Wrong On Climate Change—It is Far, Far Worse. *The Observer,* January 27.

Stern, Nicholas and Chris Taylor. 2007. Climate Change: Risks, Ethics, and the Stern review. *Science* 317 (5835): 203–4.

Sterner, Thomas and U. Martin Persson. 2008. An Even Sterner Review: Introducing Relative Prices into the Debate. *Review of Environmental Economics and Policy* 2 (1): 61–76.

Stevenson, Hayley. 2012. *Institutionalizing Unsustainability: The Paradox of Global Climate Governance.* Berkeley: University of California Press.

Stiglitz, Joseph E., Amartya Sen, and Jean-Paul Fitoussi. 2009. *Report of the Commission on the Measurement of Economic Performance and Social Progress.* New York: The New Press.

Thompson, Allen and Jeremy Bendik-Keymer. 2012. Introduction: Adapting Humanity. In Allen Thompson and Jeremy Bendik-Keymer, eds, *Ethical Adaptation to Climate Change: Human Virtues of the Future.* Cambridge, MA: MIT Press, pp. 1–23.

Thompson, Janna. 2009. *Intergenerational Justice: Rights and Responsibilities in an Intergenerational Polity.* New York: Routledge.

Throop, William M. 2012. Environmental Virtues and the Aims of Restoration. In Allen Thompson and Jeremy Bendik-Keymer, eds, *Ethical Adaptation to Climate Change: Human Virtues of the Future.* Cambridge, MA: MIT Press, pp. 47–62.

UNEP (United Nations Environment Programme). Last updated 2013. CDM Projects by Type. UNEP RISO Centre. Available at <http://www.cdmpipeline.org/cdm-projects-type.htm>.

USDA (United States Department of Agriculture). Last updated April 2013. Farmers Markets and Local Food Marketing. Available at <http://www.ams.usda.gov/AMSv1.0/farmersmarkets>.

Vanderheiden, Steve. 2008. *Atmospheric Justice.* New York: Oxford University Press.

Victor, David G. 2011. *Global Warming Gridlock: Creating More Effective Strategies for Protecting the Planet.* Cambridge: Cambridge University Press.

von Storch, Hans, Armin Bunde, and Nico Stehr. 2011. The Physical Sciences and Climate Politics. In John S. Dryzek, Richard B. Norgaard, and David Schlosberg, eds, *The Oxford Handbook of Climate Change and Society.* Oxford: Oxford University Press, pp. 113–28.

WE ACT for Environmental Justice. 2009. *Advancing Climate Justice: Transforming the Economy, Public Health & Our Environment.* Available at <http://weact.org/Portals/7/Program%20Docs/Movt_%20Bldg_/ClimateJusticeConferenceReport.pdf>.

Weart, Spencer R. 2008. *The Discovery of Global Warming,* 2nd edn. Cambridge: Cambridge University Press.

Weitzman, Martin L. 2009. On Modeling and Interpreting the Economics of Catastrophic Climate Change. *Review of Economics and Statistics* 91 (1): 1–19.

Williams, John W., Stephen T. Jackson, and John E. Kutzbach. 2007. Projected Distributions of Novel and Disappearing Climates by 2100 AD. *Proceedings of the National Academy of Sciences* 104: 5738–42.

Wilson, Edward O. 1998. *Consilience: The Unity of Knowledge.* New York: Knopf.

Index

Index